Y0-CCF-207

Energy Systems Design and Operation

About the Author

Graham Mark Tostevin is a retired consulting engineer with 50 years' experience designing and operating energy systems. He graduated in mechanical engineering from the University of Adelaide in 1953 and then served at sea in the Royal Australian Navy, where he remains on the retired list of reserve officers. He studied and worked on nuclear power plants in England and then spent four years as an industrial power plant engineer in Victoria, Australia. In 1961 he founded a consulting engineering company in South Australia.

Mark holds M.E. and Ph.D. degrees from the University of Adelaide. He is a Fellow of the Institution of Engineers, Australia, and a Member of the Institute of Marine Engineering, Science and Technology, London. He published three papers about his work on energy systems and held several positions of national leadership in his profession. He is a former member of the National Energy Research Development and Demonstration Council in Australia.

Mark was born in Adelaide in 1929. He is married to Diana, and they have a son and daughter, now adults. He is an experienced ocean-racing crewman, competing internationally for several years and twice winning the race from Sydney to Hobart. He is a life member of the Royal Ocean Racing Club, London.

Energy Systems Design and Operation

A Unified Method

G. M. Tostevin, Ph.D.

New York Chicago San Francisco
Lisbon London Madrid Mexico City
Milan New Delhi San Juan
Seoul Singapore Sydney Toronto

The McGraw·Hill Companies

Library of Congress Cataloging-in-Publication Data

Tostevin, Mark.
Energy systems design and operation : a unified method / G. M. Tostevin.
p. cm.
Includes bibliographical references and index.
ISBN 978-0-07-177291-4 (alk. paper)
1. Power (Mechanics) 2. Power resources. I. Title.
TJ163.9.T68 2012
621.042—dc23 2012005668

McGraw-Hill books are available at special quantity discounts to use as premiums and sales promotions, or for use in corporate training programs. To contact a representative, please e-mail us at bulksales@mcgraw-hill.com.

Energy Systems Design and Operation: A Unified Method

1 2 3 4 5 6 7 8 9 0 DOC/DOC 1 8 7 6 5 4 3 2

ISBN 978-0-07-177291-4
MHID 0-07-177291-X

This book is printed on acid-free paper.

Sponsoring Editor
Michael Penn

Editing Supervisor
Stephen M. Smith

Production Supervisor
Pamela A. Pelton

Acquisitions Coordinator
Bridget L. Thoreson

Project Manager
Patricia Wallenburg, TypeWriting

Copy Editor
James K. Madru

Proofreader
Paul Tyler

Indexer
Claire Splan

Art Director, Cover
Jeff Weeks

Composition
TypeWriting

Contents

Acknowledgments

This book owes its authority to P. W. Bridgman (1943), who saw and explained the nature of what we call *energy*.

I am most grateful to the Department of Mechanical Engineering at the University of Adelaide for making it possible to build on that authority and for arranging international examination of a result that has become a unified method for energy systems.

I am also grateful for the opportunities provided by several organizations in Australia to pioneer development of the unified method: CSIRO Division of Mechanical Engineering, Alcoa of Australia, Grant Spence and Associates Consulting Engineers, and Santos, Ltd.

The example "Operate" in Chapter 6 is similar to one in a book I published as a manual for clients and colleagues in Australia in 1996. The manual included some of the illustrations and a large part of the appendices of this book, but it did not contain the example computer programs that are published here for the first time.

I appreciate permission from Maney Publishing, of the United Kingdom, to use the example "Design" in Chapter 6, which is similar to one published in the *Journal of the Energy Institute* 80(2), June 2007.

I also appreciated the opportunity to speak about the essence of Chapter 1 in the *Perspective* program of the Australian Broadcasting Corporation, Radio National, on August 13, 2008.

I am indebted to all the engineers with whom I worked in a lifetime and who consequently contributed to this book, and I thank every one of the publishing team for your excellent work on the book.

My special thanks to Diana Tostevin, A.M., who, throughout her own busy life, endured and supported it all.

Introduction

This book is a contribution to energy systems engineering. It appears at a time when, all over the world, people are working on changes to energy systems, adopting new ideas and technologies to reduce emissions and costs. It tells of a way to determine the best use of energy and the best concept to adopt for an energy system in any situation. People and corporations then can deal with all kinds of energy in the same way, in any combination, for any purpose, and always for the best outcome. Everyone concerned with the use of energy should read this, especially the directors, managers, and instructors of those who conceive, design, and operate energy systems.

In the main, people have learned how to design and operate energy systems as if by apprenticeship to previous generations. Engineering schools and books tell them about the machines and equipment of energy systems, options and alternatives for design, and ways to optimize particular kinds of systems. Not yet, it seems, are they told how to conceive, design, and operate the best energy system in a given situation. This book does so. It presents a formal, unified, and universal method for all such work. This method began with an insight of Professor Bridgman—Nobel laureate in physics and faculty member at Harvard University—during the 1940s, but it required modern computers to bring the method into practice.

This book centers on the functions of the machines of energy systems rather than on their theory, design, and construction. Thus it repeats little of the content of other works on such subjects as boilers, turbines, engines, motors, generators, refrigerators, and heat exchangers, although it implicitly includes all such information.

The text of this book is concise so that readers can quickly see the essence of the unified method. Seven chapters tell of the

- Reason for a unified method
- Gathering and processing of information for energy systems, a necessary task whatever the method
- Formality required to design and operate energy systems, including those with energy storage
- Extension to the formality so that it becomes unified for all energy systems
- Computing required for such a method
- Use of the method in practice, demonstrating optimal operation of an industrial power station and design of the optimal energy system for an office building
- Answers to typical questions that arose during research, development, and application of the method

Supporting material includes, in the appendices, the algorithm and functional specification of the computer program for the unified method, and explanations of current techniques for application of the method; and, as a download from the website associated with this book, www.mhprofessional.com/TostevinEnergySystems, a collection of computer software that includes example computer programs with their source code.

After reading the text, decide whether there is a place for the unified method in your own corporation or field of work. If so, use the supporting material as starting points for the people who will put the method into practice. Some will create the computer program for use on all your energy projects; others will maintain the unified database.

The principles of the unified method and the understanding they convey of the nature of energy systems will endure. Yet the techniques of application will continue to evolve in the future as more people use and contribute to the method. Global concern with the use of energy seems to ensure that this will happen.

CHAPTER 1

A Reason

To Adopt a Unified Method for Energy Systems

Let us talk about the common task and foremost problem for people who conceive, design, and operate energy systems: Of all the available options and alternatives, always choose the best.

By *energy systems*, I mean those which provide electricity, steam, propulsion, heating, cooling, and other services for public power supplies, process industries, ships, factories, and buildings.

People have undertaken this task for centuries. A farmer decides to use a wood-fired heater and storage tank for hot water with a wind generator and batteries for electricity. The owner of a power station installs coal-fired boilers and steam turbo generators. A government approves plans for a hydroelectric and pumped storage scheme. An investor supports the building of a nuclear power plant. A ship owner installs diesel electric propulsion. The owner of a building uses absorption refrigeration for air conditioning. Each is a unique energy system, chosen from a variety of possibilities. In each case, people will have gathered information, recalled from experience, and evaluated several options before deciding on the energy system to adopt.

The task became a problem as it became more difficult. Energy once seemed plentiful and inexpensive, and the options for energy systems were relatively few and simple. In the 1970s, however, the price of oil soared. People working with energy systems faced calls at once to reduce fuel consumptions and costs, improve efficiencies, use different sources of energy, and employ new technologies. More recently, the calls have intensified to reduce emissions.

The difficulty of deciding the best energy system grows quickly as the number of options increases, especially when they are unfamiliar. Each different combination of options represents an alternative energy system, of which there may be many to consider for a large project. Each will have different attributes and consequences—technical, financial, environmental, and social—so more people become involved. Bankers, regulators, and environmentalists now usually join owners, designers, and operators to make decisions about energy systems. All these people need to understand and trust the method used to evaluate alternatives and to present them for decision. They need to see that the method

- Deals rigorously and equally with each alternative
- Uniformly compares the attributes, costs, and consequences of any one method with those of another
- Identifies the best energy system for any chosen criteria
- Provides a reliable specification for the design and operation of the energy system they decide to adopt

All this requires a more formal and systematic method than might have been sufficient in the past for a few simple options. Far from being difficult, however, when we formalize the method for this, we find that it becomes the same, or unified, for all energy systems.

Global navigation and communication are examples of unified methods, and they begin with an organization of information in universal terms. For example, while still preserving the individuality of each ship and its voyage, parameters of latitude, longitude, direction, distance, and time mean the same things to all seamen and their systems worldwide.

Now here's the point: It is in the nature of energy systems that they too can be expressed in universal terms. This offers the key to a unified method, one that equips us to deal with all energy systems in the same way, universally, and especially to decide the best of all the options and alternatives in any given situation.

A unified method also brings immediate practical benefits to the people who work on energy systems: uniform language and communication, common training and computer programs, inbuilt optimization, coordinated research and development, and universal recognition.

It would seem to be a huge task to establish such a method globally or even nationally. However, it is a characteristic of unified methods that they can begin on many fronts and in small ways, all inherently compatible with each other. As they evolve, more and more people, corporations, and nations can participate and collaborate. Let's begin.

CHAPTER 2

The Task

Of Gathering and Processing Information for Energy Systems

People of many vocations work on energy systems from concept through design, construction, years of operation, and their eventual replacement or removal. Everyone shares responsibility for the success of an energy system, and their common thread is information. At every stage of their work and in some way, they gather and process information in order to make the right decisions, and this book tells of a unified way that is the same for all energy systems.

Concept means a particular kind of energy system, distinguished by its combination of sources of energy, processes, plant, and demands. *Design* means producing the engineering plans and specifications needed to build and commission an energy system. *Operation* means the day-to-day running and management of an energy system after it is built and commissioned.

Gather Information

Work on an energy system begins with people researching and then assembling the necessary information, ready for processing. The information required for concept and design includes the following:

Location. Geographic position and altitude, meteorological records (air temperatures and pressures, wind speed and direction, solar radiation), records of the qualities and levels of local water supplies, statutory requirements, and permissible limits for the effects of an energy system on the environment (emissions, exhausts, effluents) and on society (noise, traffic, vibration).

Demands. Form and qualities (electricity, steam, propulsion, refrigeration, hot water, chilled water), expected values (average, maxima, and minima at different times of the day, week, and month of a typical period of operation), and economic worth.

Sources of energy. Form and qualities (coal, solar, oil, wind, gas, nuclear, hydro), expected availabilities and limits of supply at different times of the day and days of the year, expected costs (tariffs, conditions of contract), and requirements for waste disposal.

Plant. Details of machines and equipment available for use in the energy system (names, makes, sizes, efficiencies, controls, costs, space, and weight), requirements for their auxiliary plant and services and for their installation (transport, foundations, structures, pipes, ducts, and cables).

Interconnections with other energy systems. Tie to the public electricity grid, for example.

For concept and design, equally complete information is required for each option and alternative.

For operating an existing energy system, similar information is required, although mostly then with current rather than expected values and with fewer options and alternatives.

All the preceding is *given information*, the raw material from which the subsequent processing will produce its results. The given information has to be complete and reliable for the results to be reliable. Careful acquisition and maintenance of the given information is an ongoing part of the task.

A unified method standardizes the files and utility computer programs required to contain and manage the given information. Those files and programs provide a place for everything, and when everything is in place, the information is complete.

Process Information

Processing the given information follows a sequence that can be the same for all energy systems: calculation, integration, evaluation, synthesis, optimization, and conveying the results.

Calculation begins with a set of values for current demands. It then works through a particular concept for an energy system to determine its least use of plant and least supplies from sources of energy for the *current time*.

Values of demands vary with time. Unless you are dealing with just the current operation of an energy system, the processing therefore repeats the calculation, for each minute or hour, for example, through an *extended period*.

Integration accumulates the results of such calculations to determine the use of plant and the consumptions of sources of energy for whatever extended period is required for evaluation, such as a day, a week, a month, or a year.

Evaluation, at the end of the extended period, calculates the values of certain criteria, such as efficiency, costs, and emissions, together with space, weight, appearance, and construction time when designing a new plant, by which the processing and the people concerned can compare the attributes and consequences of any one energy system with those of another.

Synthesis imposes different sets of options and alternatives systematically, in incremental steps, to repeat the calculation, integration, and evaluation for each different energy system that is possible within the scope of the given information. The processing records the identity of each step together with its values for evaluation criteria.

Optimization first requires a choice of one or more evaluation criteria to form an *objective* by which to identify the best energy system. After completing all the steps of synthesis, the processing finds the step that most closely meets that objective and then resynthesizes the detail of the energy system at that step.

Conveying the results means sorting and analyzing all the different energy systems according to their values for evaluation criteria and then, on request, resynthesizing and displaying the detail of any one of them.

Calculation, integration, and evaluation are straightforward engineering computations, for which a computer quickly produces the results for each step of synthesis. Optimization is also straightforward because it only has to find a particular step in a table of values for evaluation criteria and then resynthesize.

Synthesis is a creative (building block) procedure, rather than an analytical procedure. The time taken for processing therefore depends on the sizes, and hence the number, of incremental steps of synthesis. Those steps always must span the whole range of concepts, options, and alternatives within the scope of the given information. However, we may choose the increments, to be coarse or fine, for example, so that the time taken is appropriate for the required precision of the results.

A unified method standardizes the computer program for all such processing.

CHAPTER 3

Formalize

Engineering Principles for the Design and Operation of Energy Systems, Including Those with Energy Storage

Imagine for the time being that we have already decided the concept for a new energy system and have chosen from all the options and alternatives. Two parts of the processing of information, namely, synthesis and optimization, are then virtually inactive. However, we still have to conduct the other parts of processing—calculation, integration, and evaluation—just to design the energy system. This chapter describes a formal procedure for this. In essence, it expresses what designers of energy systems already know, but in a way here that is the same for all energy systems.

Formality begins by recognizing the dual character of an energy system. Figure 3.1 shows the two parts of an energy system. Part *A* consists of all natural processes. Part *B* is the manufactured plant. They are two different things. Processes disappear, for example, if their source of energy stops, yet the plant is still there, cold and silent. The formal procedure deals with both parts of an energy system concurrently.

The processes are those of conversion, exchange, storage, and transmission of energy. Certain sequences of such processes are required for particular sources of energy to meet particular demands. The processes in a steam turbine, generator, and transformer, for example, enable a source of steam to meet a demand for electricity.

We define the *extent* of an energy system—and the scope of our work on it—by the notion of a boundary. Sources of energy originate from the region outside the boundary, and the energy system delivers to demands from that region.

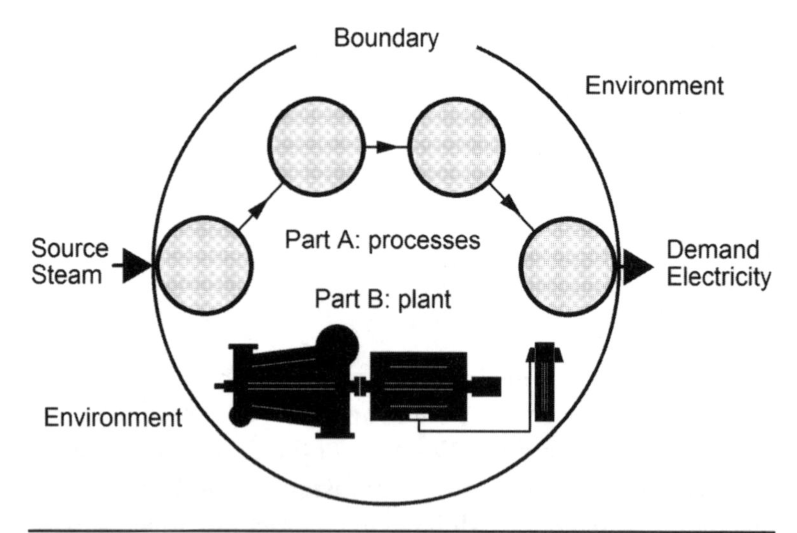

Figure 3.1 Two parts of an energy system. (*A*) Sources of energy supply certain sequences of processes that deliver energy to certain demands, in the required forms, at the required rates, and for a particular time. (*B*) An assembly of machines, equipment, pipes, ducts, and cables designed, built, and operated to contain and control the processes.

Physical, biologic, and sociologic conditions in the region outside the boundary describe the environment of an energy system. Some processes of an energy system, such as combustion, can affect the environment, and conditions in the environment, such as the temperature of the air, can affect some of the processes.

Processes: First Law

We also can apply the idea of a boundary to either a single process or a group of associated processes inside an energy system. Each small boundary then forms a *cell* around those processes.

All the cells are similar, in the sense that each is accompanied by *transfers of energy*, of whatever form—chemical, thermal, mechanical, or electrical—all accountable to the *first law of thermodynamics*. Each cell has a needed transfer of energy and a supplied transfer of energy, and the directions of these transfers may be into or out of a cell depending on the nature of its processes. A heating process, for example, has a needed transfer out, whereas a cooling process has a needed transfer in.

Many cells also have a residual transfer of energy, such as heat from an engine's exhaust. Most cells have parasitic transfers of energy, such as losses and gains of heat and fluids to and from the environment (Figure 3.2).

Each transfer of energy has a numerical value. Setting aside the effects of time and energy storage for the moment, the algebraic sum of the transfers of energy into and out of each cell is zero. In other words, the transfers of energy for each cell are balanced.

Transfers of energy connect the cells. The supplied transfer of one is the needed transfer of another. A network of cells with continuity of transfers of energy between sources of energy and demands, then, represents a *cellular view* of an energy system (Figure 3.3).

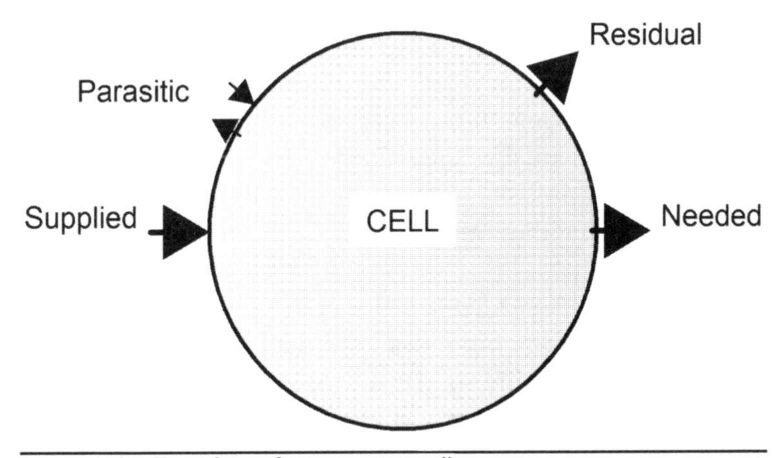

FIGURE 3.2 Transfers of energy at a cell.

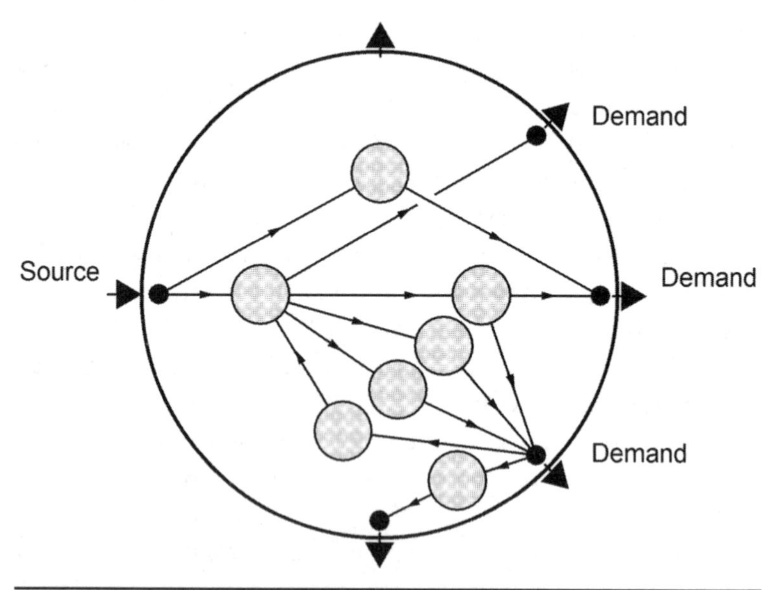

Figure 3.3 Cellular view of an energy system.

Sources of energy and demands, respectively, provide the supplied and needed transfers of energy for cells at the boundary of an energy system. Residual transfers of energy may discharge to the environment at the boundary, or they may become supplied transfers to other cells when, for example, recovering heat from an engine's exhaust. Just as they are for each cell, the transfers of energy for the whole energy system are balanced.

This cellular view of an energy system provides a formal structure for calculating the values of its transfers of energy. Beginning with one or more given demands, the calculation works upstream through the cells in sequence toward the sources of energy.

The value of a needed transfer of energy provides a measure of the *size* of the processes in the cell. Other things being equal, for example, the size of the processes would double if the rate of needed transfer of energy doubled. However, other things are rarely equal, and a change in the size of processes depends on their physical, chemical, and thermodynamic detail. This detail includes measures of the properties of the materials (usually working fluids) in which the processes are occurring, such as their temperatures, pressures, densities, and specific energies. It also includes a measure of the efficiency of any conversion of energy in the cell, and this determines the relation between the values of its needed, supplied, and residual transfers of energy. We can imagine that each cell "carries" all this information about the size and detail of its processes.

A cell also can carry information about the costs of its processes. Most processes occur naturally so that, excluding their plant, they are free of cost. However, costs usually apply to supplies from sources of energy such as fuels and electricity and therefore to the processes that use them.

Processes: Second Law

The *second law of thermodynamics* rules the directions of transfers of energy and therefore the sequences of processes and connections between the cells. For example, heat may transfer out of one cell into another with processes at sufficiently low temperatures but not directly into one with processes at higher temperatures. The latter connection of cells fails *thermodynamically*, and the sequence of processes that it would otherwise represent is not feasible.

Further, the transfers of energy out of some cells might be physically unable to connect with others. For example, heat out of a combustion process cannot usefully connect into an electromagnetic process, and such a sequence fails *physically*.

Imagine also that the numerical value of a transfer of energy out of one cell is insufficient to maintain the processes in another connected cell, and the sequence fails *numerically*.

The formal procedure admits all sequences of processes that can help to meet required demands from available sources of energy, but it discards at once all sequences for which the continuity of transfers of energy would or does fail. Computer programs deal with this on a logical basis: that a particular sequence of processes—or indeed a whole energy system—is either feasible or not.

Plant

Physically, it will be a part of the plant, rather than a cell, that contains a particular process or group of processes. However, we can place the boundaries of cells at will. We therefore assign cells so that they correspond with the processes in a particular *section of plant*. The cellular view of an energy system then matches the physical *configuration* of the whole plant (Figure 3.4). Chapter 6 also shows Figures 3.3 and 3.4 in more detail.

A section of plant may comprise just a single unit. Usually, however, it includes several units, such as five boilers, each able to contain a similar group of processes. Just as a cell can represent a process or group of associated processes, it also can represent either a single *unit of plant* or a whole section of plant.

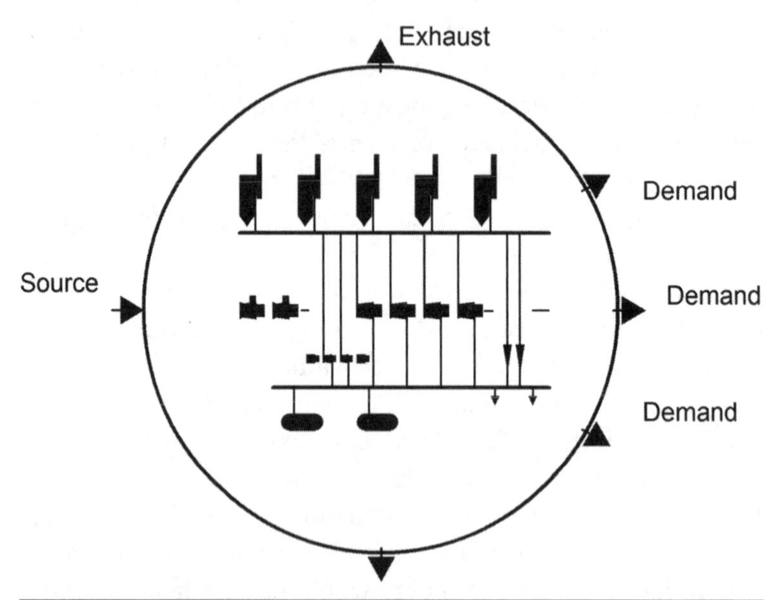

FIGURE 3.4 Configuration of a plant.

Each section of plant has to be big enough to contain and control the sizes of the processes associated with the maximum demands on the energy system during certain extreme conditions. It also has to be adaptable enough to contain and control whatever lesser sizes of processes are associated, from time to time, with current demands and conditions. Designers of energy systems determine and specify the requirements for each section and unit of plant accordingly.

Imagine, then, that the cells of an energy system, already carrying information about their processes, also carry information about their plant. This will include details of the make, model, rating, current load, and efficiency of each unit of plant, its conditions and limits of operation, controls, auxiliary plant and services, space, weight, and costs.

The *rating* of a unit of plant corresponds to the maximum size of the processes it can contain and therefore to the maximum rate of needed transfer of energy at certain standard conditions. The *load* on a unit corresponds to the size of its processes and rate of needed transfer of energy at current conditions.

The measure of efficiency for a unit of plant usually combines the natural effects of any conversion of energy in its processes with the mechanical effects of its particular design and manufacture. A mathematical expression or numerical table can provide the value for that efficiency (or coefficient of performance) for a unit over a range of load and conditions. Each unit requires such information, which can be expressed initially as a performance curve for the unit at certain standard conditions.

Information about the expense of owning and using a unit of plant includes its capital cost ($), annual fixed cost ($ per year), and running cost ($ per hour of operation).

Each cell of an energy system can carry all such information for the section or unit of plant that it represents. The cellular view then provides a formal structure not only for calculating the processes of an energy system but also for determining the requirements of its plant and finally for evaluating them all together.

Calculations for energy systems also have to account for parasitic transfers of energy to and from the cells. Some values will depend on differences between the conditions in the cell and the environment, such as transfers of heat to or from refrigerant pipes. Others will be associated with mechanical losses, such as steam leaks from a plant.

Many kinds of plant require the installation of auxiliaries, such as pumps and fans. *Major auxiliaries*, of significant size, may be included in the cellular view of an energy system as a separate section of plant in their own right. However, it is often practical to account for the effects of *minor auxiliaries* as a notional auxiliary transfer of energy associated with a particular cell. Calculations for energy systems then account for transfers of energy required by minor auxiliaries—and indeed, for any other such services—as *internal demands*. Some units of plant may have *packaged auxiliaries*, the effects of which will be included in the stated net performance of the unit (see "Auxiliaries" in Appendix C).

Time

The sizes of processes in the cells of an energy system would be constant if the demands, sources of energy, and conditions in the cells and the environment remained constant. However, the values of demands typically vary from moment to moment, the availabilities and qualities of sources of energy often vary, and the conditions in the environment are forever changing.

The formal procedure deals with these variations by declaring a *unit time* for which all the conditions in and affecting an energy system are notionally constant. A unit time of one second may be necessary for experimental work, but one hour (or half an hour) is usually appropriate for industrial and commercial energy systems. The processing calculates its results independently for each unit time and integrates them through whatever extended period is required for evaluating or managing the operation of an energy system.

With the dimension of time included, measures for demands, supplies from sources of energy, transfers of energy, sizes of processes, and ratings of plant all can be expressed as *rates* of transfer of energy in universal terms such as megawatts (MW) or British thermal units per hour (Btu/h). The formal procedure maintains numerical values of *conversion factors* to change such rates to physical values when required, such as the flow of working fluids for hydraulic and aerodynamic calculations.

Energy Storage

Variations in the values of demands or supplies from sources of energy provide opportunities—or even make it necessary—to include storage in an energy system. Optimal design and operation make the best use of storage if it is available.

Physically, storage can be provided as a separate section of plant in forms such as hydroelectric dams, hot-water tanks, ice tanks, and electric batteries. In the cellular view of an energy system, storage then appears as an additional cell connected to those of other processes (Figure 3.5). As with other cells, those

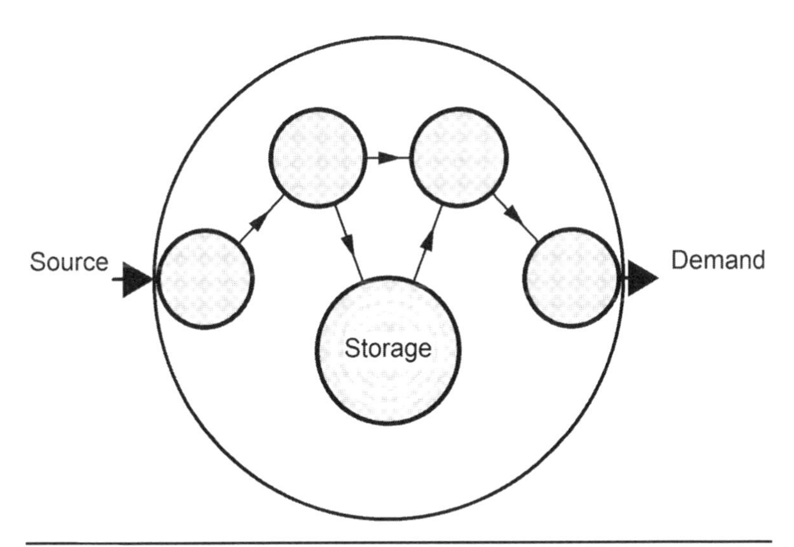

FIGURE 3.5 Cellular view of an energy system with storage.

for energy storage carry information about both their processes and the plant.

The formal procedure accounts for the transfer of energy into and out of storage for each unit time, as it does for other cells. However, differences arise with the calculation and integration of energy systems with storage as follows:

1. Storage means the accumulation of supplied transfers of energy *in* at some times, followed by their release as needed transfers *out* at subsequent times, all within the period of a *store cycle*. For such a period, the first law of thermodynamics requires the sum of supplied transfers into storage to equal the sum of needed transfers out. Calculations for energy systems with storage therefore always require integration through a period of at least one store cycle.
2. The purpose of storage is to provide needed transfers of energy out at times when other connected processes are unable to do so. During design, the processing needs to know the values of the needed transfers out before it can calculate the required transfers into storage at some previous times. Therefore, when storage is included in an energy system, the processing integrates in reverse order of time. Such a reversal does not affect the results of calculations and integrations for other parts of energy systems that are not associated with storage.
3. The required physical size of storage then evolves as the processing integrates through each store cycle (see Appendix A and "Energy Storage" in Appendix C).

Design and Operation

The formal procedure for design of an energy system necessarily includes simulation of its operation through a period of varying demands, such as a typical year. It is inadequate to design an energy system for just its maximum demands because the effects of part load operations and variations in the sources of energy and the environment usually dominate its management and overall costs.

Although the tasks of designers and operators of energy systems are physically different, they are similar as far as their use of information is concerned. Both have to consider the sequences of processes that are required from time to time to meet certain demands from available sources of energy and decide the best. Designers decide from sections and units of plant that can contain the necessary processes and are available commercially, and operators decide from sections and units of plant that are already installed and available operationally. Both designers and operators employ anticipated schedules for demands, for the availabilities of sources of energy, and for the conditions of the environment. In essence, the formal procedure and the processing of information are the same for both design and operation.

CHAPTER 4

Unify

Extend the Formality So That It Becomes Unified for the Concept, Design, and Operation of All Energy Systems

Until now, the formal procedure has been dealing with just a single concept for an energy system. This may have been sufficient in the past, when people could determine the best concept from experience. However, in an effort to reduce costs and emissions and use new energy technologies, it is now usually necessary to consider several different concepts, the best of which remains open to question.

This chapter tells you how to extend the formal procedure for this to evaluate each possible concept within the scope of the given information and determine the best. At the outset, it seems self-evident that this will require a form of information processing that is the same for all concepts.

Universal Organization

The formal procedure provides the foundation for such an extension because it already applies to all kinds of energy systems and deals with information about transfers of energy and ratings of plants in universal terms. It only remains to establish universal terms for the information carried by the cells of energy systems, and we discover this in the nature of the processes themselves.

P. W. Bridgman (1943, p. 84) once explained that "we should talk about the energy of a body or system only when it is brought by some process from an initial to a final state, and should not speak of the energy of the initial or the final state by itself." He named the pair of initial and final states as a *state-couple*.

This insight points to the existence of a universal structure for information about the processes of energy systems. In the cellular view, for instance, it means that each cell will contain one or more state-couples and that, as far as its transfers of energy are concerned, we can express all the information we need about its processes in terms of just their initial and final states. The formal procedure already includes equivalent information, although arranged arbitrarily. When arranged in terms of state-couples, however, the pattern of information becomes consistent for all cells, and this is the essence of unification.

For energy systems of the physical sciences, we can express the information for each cell in terms of initial and final temperatures, pressures, and other properties of materials and the thermodynamic or physical path associated with the change of state. The rates of needed, supplied, and residual transfers of energy will be a function of that information and the size of the processes. When we arrange the content of the cells in such a manner, we find a universal organization of information that is the same for all cells (Tostevin and Luxton, 1979).

One way of visualizing such an organization is to imagine that we put the details of each cell into a set of containers (Figure 4.1).

We need containers for

Size. The size of the processes, based on the value of the rate of needed transfer of energy from the cell.

Temperatures. Initial and final temperatures of materials associated with the processes.

Pressures. Initial and final pressures of those materials, particularly working fluids.

Materials. Initial and final properties of the materials at those temperatures and pressures, especially their specific (internal, gravitational, kinetic) energies and densities.

Efficiency. A mathematical expression or table to provide a value for the efficiency of any conversion of energy in the

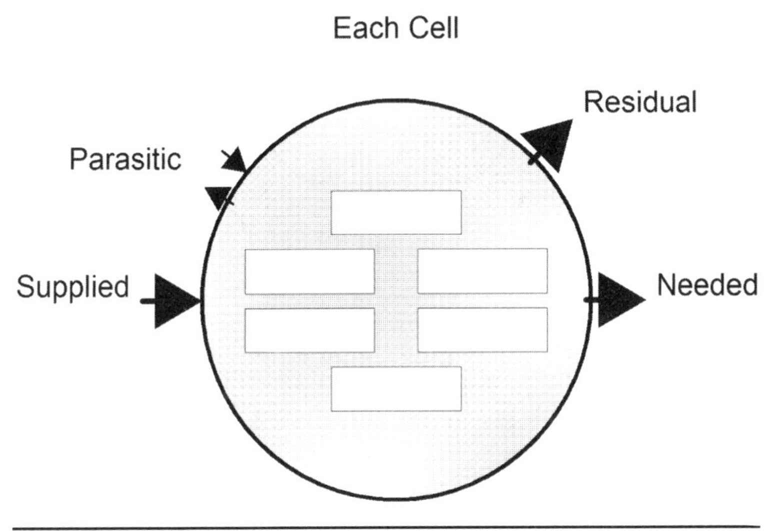

FIGURE 4.1 Containers of information in each cell.

processes, representing the effects of their thermodynamic and physical paths.

Just as a cell carries information about both processes and plant, each container also can carry the relevant information for both. The size container, for example, can carry values not only for the current load on the plant but also its permissible limits of load. The temperature container can carry values for the limits of temperature for operating the plant. The efficiency container can carry a measure that accounts for the combined effect of both the natural processes and the manufactured plant.

Finally, for evaluating the economics of energy systems, we need a sixth container for

Costs. Information about the costs of the processes and plant represented by the cell.

All cells do not necessarily require all such details, but if they do, there is a place for them. Cells containing electromagnetic processes, for example, do not normally need information about pressures. Chapter 5 assigns names for the containers.

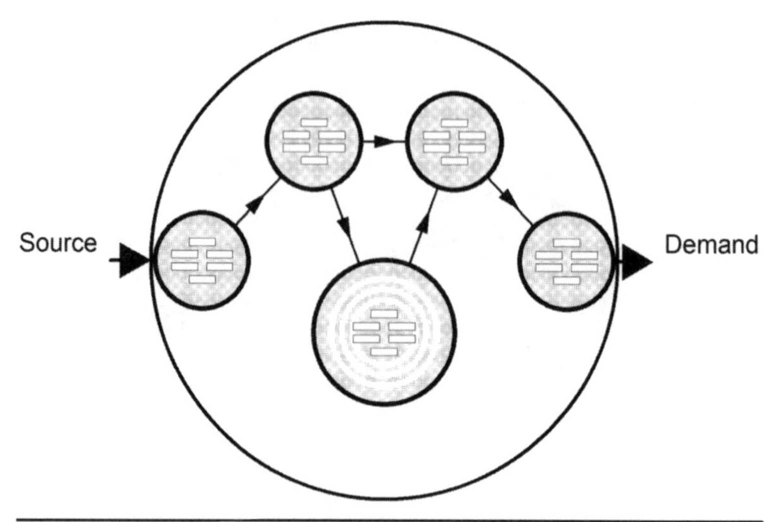

FIGURE 4.2 Containers of information in all cells.

The information carried by the cells is the same as that already required by the formal procedure, but now that information is organized in a universal way. The arrangement of containers and their information is the same for all cells (Figure 4.2).

With such an organization of information, the formal procedure becomes a unified method that deals with all parts of energy systems in the same way in terms of transfers of energy, cells, containers, and their content.

Extend the Formality

As an extension to the formal procedure, the unified method simply will acquire all the given information in universal terms and then conduct the processing in those terms. In addition, it now will include the two parts of processing previously inactive in the formal procedure, namely, synthesis and optimization.

The unified method synthesizes and evaluates the design and operation of each different alternative for an energy system in turn. It evaluates them all and resynthesizes the details of any one on request. It identifies the best energy system for any chosen objective. Appendices A and B describe the details of this processing.

An overall view of the unified method is that it provides sets of cells and containers—a universal organization of data for energy systems—into which people always know where and how to put the given information. Computer programs then know where to find that information and what to do with it for each part and the whole of all energy systems.

Uniting Energy Systems

Finding the best of several concepts does not necessarily mean that an energy system also will be the best if it has to operate in conjunction with other systems. Just as each different concept for one system requires evaluation, so does each different union of multiple energy systems.

Energy systems may be united in many different ways. One way is to interconnect their demands; connecting individual power stations to the public electricity grid over a wide area, for example, can provide major operational and commercial benefits. Interconnecting the processes of nearby energy systems is another way; residual heat from one system, for example, might become a source of energy for another. Yet another way is to aggregate several small, similar energy systems into one large system; a district heating scheme, for example, may replace hundreds of small residential fuel-fired heaters.

The unified method requires no change in principle to deal with the union of energy systems; and it simplifies and speeds the work of examining interconnections of different systems. We gather the information about each system in universal terms; the processing synthesizes all possible unions, then it identifies and resynthesizes the best for any given criteria.

Opportunities and incentives for uniting energy systems abound, and they increase as our view expands from a local to a corporate to even a national scope. Each practical opportunity usually deserves evaluation, even if only to discover the extent of any benefits it might offer.

Another view is that all the energy systems in the world are already working in some vast union and that it has become a global imperative to reduce emissions by changing the mix and using different sources of energy and different technologies. Such an ambition also implies a need to unite the efforts of everyone concerned with energy systems—something that a unified method of work and communication may well help to achieve.

Computing Power

As the scope for uniting energy systems expands, however, the unified method requires higher computing power. It will have greater capacity to address more information, faster speed of processing, and support from advanced methods of programming.

Evaluating the union of energy systems on a wide scale and deciding the best therefore is a goal for the future. Yet it is the kind of goal already achieved for global navigation and worldwide corporate operations. The unified method, with its organization and processing of information in universal terms, leads to such a goal. Meanwhile, in 2012, even a notebook computer is sufficient to design and operate the union of energy systems on a common corporate scale.

CHAPTER 5

Compute

Requirements for Computing with a Unified Method

Processing information for energy systems was once a task for slide rules and mechanical calculators, but we now use computers. This requires a system of symbolic names to carry the given and computed information, a database in which to assemble the given information, and finally, a computer program written in those names to conduct the processing. The unified method requires and provides for all this in a way that is the same for all energy systems.

Symbolic Names

The cellular view of an energy system, with its containers of information, offers the basis for a system of symbolic names. Imagine, for example, that the symbol (Q) represents the set of values for the current rates of needed, supplied, and residual transfers of energy at each cell and that symbols SIZE, TT, PP, MM, EFFY, and VV represent and name the containers in each cell for information about sizes, temperatures, pressures, materials, efficiency, and costs, respectively.

Deferring for a moment the subjects of costs and parasitic transfers of energy, the expression

$$(Q) = f\,(\text{SIZE, TT, PP, MM, EFFY})$$

then represents a universal organization of information throughout the cellular view of an energy system. It is a concise way of saying that for a unit time in which conditions are steady, if we know the information in the containers of a cell, we can determine the values for its rates of transfer of energy. This relation rules the calculations for transfers of energy in the design and operation of energy systems.

Returning to the subjects deferred, the processing calculates the costs associated with each cell from the information in VV in such a way that the costs for a whole energy system are the sum of those of all cells. Parasitic transfers of energy often depend on the conditions in the environment or in other parts of the energy system, so the processing calculates them for each cell independently and applies them as adjustments to supplied or needed transfers of energy.

The preceding symbols for transfers of energy and containers provide a basis for a universal system of names to carry the detailed information for the cells of energy systems—as in the following examples:

Transfers of energy (Q). Each cell requires a set of symbolic names, such as QN, QS, QR, and QP, to carry values for its rates of needed, supplied, residual, and parasitic transfers of energy. An energy system usually includes several sections of plant, each with several units. We can index the names numerically to distinguish one cell from another. The names then are QN[NN] for a section of plant of index NN and QNU[NN][NU] for a unit of index NU in that section (Figure 5.1). Similarly, the names QA and QAU carry values for the rates of auxiliary transfers of energy for any minor auxiliaries associated with a cell.

Containers. In essence, the information in the first five containers (SIZE, TT, PP, MM, EFFY) only has to describe the identity, sizes, and state-couples of the processes and plant represented by a cell, in terms of both current values and limits of value. It is not necessary to include other technical detail. Container VV merely carries values for costs and other data required for evaluations.

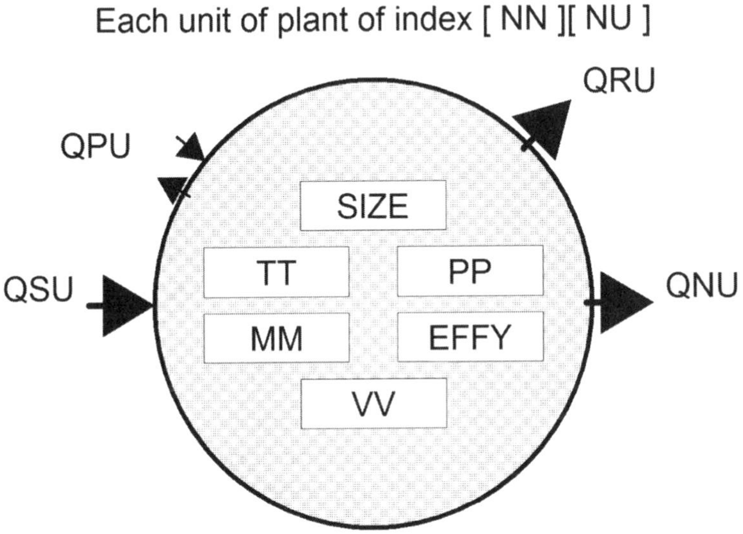

Figure 5.1 Symbolic names at a cell for a unit of plant.

For example, the SIZE container carries its detailed information in symbolic names beginning with SZ, such as SZHI for the maximum permissible load on the plant represented by a cell. The TT container carries information in names beginning with T, such as TI for the initial temperature of a process represented by a cell. The PP container carries information in names beginning with P, such as PFMAX for the maximum permissible pressure for the plant represented by a cell. In the MM container, symbol WM carries the literal name of a material. The EFFY container carries the information needed to obtain the value of efficiency for a cell at certain conditions and then places that value in the same symbolic name, EFFY. The VV container carries information in names beginning with V, such as VC for the capital cost of the plant represented by the cell (see "Containers" in Appendix C).

As with the symbolic names for transfers of energy (Q), we index those in containers to distinguish one cell from another using the same indices, NN and NU. The name SZOP[NN][NU] then carries the value of the load on unit NU in section of plant NN.

Sources of energy and demands. At the boundary of an energy system, symbolic names QE and QD, respectively, carry the rates of transfer of energy associated with sources of energy and demands, indexed when necessary with symbols NE and ND, respectively, to distinguish one kind of source or demand from another.

Time. Unless dealing with just the current operation of an energy system, we have to index several symbolic names, such as QE and QD, so that they carry the value associated with each unit time. Indices IM, ID, and IH define the month, day, and hour, respectively. The name QD[ND][IM][ID][IH] then carries the value of a demand of index ND for a particular hour of a year. Names beginning with E, such as EQD[ND], carry values that have been integrated through an extended period.

Environment. Information about the environment also requires a set of symbolic names, such as TADB and PA, indexed for the unit time, to carry values for the dry-bulb temperature and the barometric pressure of ambient air.

Standard conditions. Each energy system requires a statement of certain standard conditions as a basis for defining the ratings of its plant. The processing refers to such standard values throughout its calculations to adjust the ratings and performance of the plant for current conditions. Appending character Z to a symbolic name indicates that it carries information associated with standard conditions. The name SZHZ in container SIZE, for example, carries the value for the maximum continuous rating of a unit of plant at standard conditions, whereas TADBZ and PAZ carry values for the standard temperature and pressure of ambient air (see "Standard Conditions and Constants" in Appendix C).

The preceding are just examples, but they indicate the way that people can construct a universal system of symbolic names for the unified method and its computer programs. Those who wish to coordinate their work with that of others will use a common system of names.

Computer programs use the symbolic names both to carry values of information for energy systems and to express the algebraic calculations for their processing. For example, at a cell representing a nonrefrigerating section of plant of index NN, for a given value of its rate of needed transfer of energy, the following lines of source code produce values for its supplied and residual rates of transfer of energy:

```
QS[NN] = QN[NN] / EFFY[NN];
QR[NN] = QS[NN] - QN[NN];
```

A computer program for the unified method is a long but highly organized sequence of such code.

Another effect of such a system of symbolic names is that the information required for all the different kinds of plants—boilers, turbines, condensers, engines, motors, generators, heat exchangers, transformers, compressors, solar collectors, energy storage, pumps, and fans—will be expressed in the same universal terms. Manufacturers may present data for their plant in such terms; otherwise, we have to convert it, and this becomes part of the task of gathering information for an energy system.

Database Files

The unified method requires three kinds of computer files to hold its given information.

1. A *time-base file* that is specific to the location and purpose of each energy system contains information about sources of energy, demands, and conditions of the environment for each unit time through a certain period of

operation. Both maximum and average values may be required; the former rule the size of the plant, whereas the latter rule the consumptions of energy. For concept and design, the information in the time-base file has to provide for every option and alternative being considered. However, once an energy system is built and operating, the time-base file—if used at all—provides just the information needed for scheduling future operations.

2. A *machinery file* lists the identity and detailed information for each different kind of plant that is available for use in an energy system, all expressed in universal terms. For concept and initial design, the file contains this as *generalized information*, typical for each section of plant. For detailed design, the file contains *specific information* for particular makes and units of plants. For operating an existing energy system, the file contains *current information* about the units of plants that are actually installed (see "Generalized and Specific Information" in Appendix C).

   The machinery file also includes definitions of the standard conditions and values of certain constants that apply throughout an energy system (see "Standardized Conditions and Constants" in Appendix C).

   The demonstrations in this book use a single *performance curve* to describe the value of efficiency or coefficient of performance of a unit of plant over a range of loads at standard conditions. The machinery file then includes the values of a set of ABCD coefficients for each unit to represent a mathematical approximation of that curve (see "Plant Performance Curves" in Appendix C).

3. A *tariff file* is required for energy systems exposed to a variety of commercial offers for the supply of sources of energy such as gas and electricity. For each step of synthesis, the processing evaluates the use of each available tariff and selects the best.

Units of measurement have to be consistent throughout the database files and the computer program. We may choose these units at the outset of a project and then not change them except for strictly local use, such as a visual display of information in alternative forms. Our choice of units begins with the measure for a rate of transfer of energy, such as megajoules per hour (MJ/h) or British thermal units per hour (Btu/h). Values for the information in containers then must conform to the same system of measurement, metric or English.

Information in the database files becomes a valuable corporate resource, and the files require careful maintenance to keep them up to date. Utility computer programs help people to create and maintain these files and to translate manufacturers' data into universal terms. The demonstrations in this book provide examples (see Appendix D).

Computer Program

With all the given information in universal form, the arrangement of the computer program for the unified method is essentially the same for all energy systems.

A separate part of the program conducts each principal part of the processing—synthesize, evaluate, integrate, and calculate. In addition, an executive part of the program manages the processing and conducts the resynthesis and optimization (Figure 5.2).

Appendix A describes the algorithm for such a computer program, and Appendix B provides a functional specification for the program.

The work of producing such a program is a permanent investment. Much of a program built for one energy system can be used for others. We can begin with a small program and expand it progressively because the new pieces always will fit with the old. A program used for the design of an energy system can manage and optimize its operation subsequently.

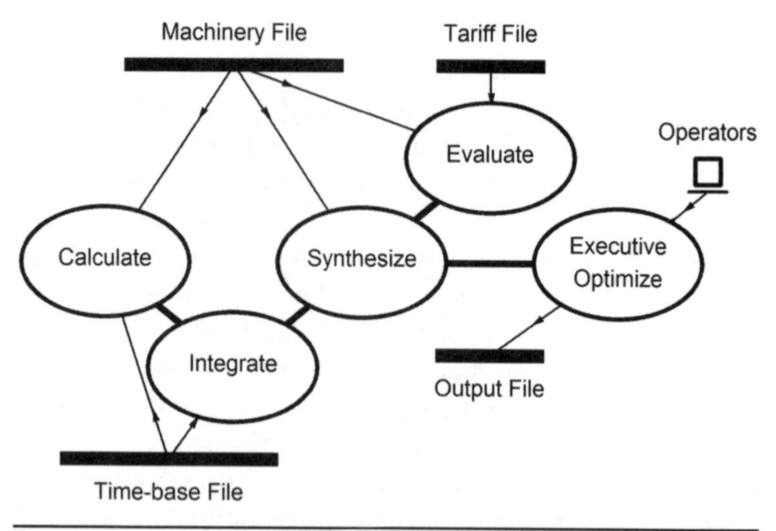

Figure 5.2 Arrangement of the computer program.

Most computer programs for the unified method require some degree of detailing to suit their particular application as, for example, in the programs for the demonstrations in this book. The detailing can meet any required corporate standards, although care is needed to maintain and not degrade the universal organization of information.

At all times, a computer program for the unified method should satisfy the following principal tests:

- It should synthesize the detail and correctly specify optimal values for the design and operation of all concepts for an energy system that are possible within the scope of the given information.
- It should do so over the whole practical range of deliveries to demands, availabilities of plant, and supplies from sources of energy.

For operating an existing energy system, a computer program can make an initial *static calculation* based on the current loads on the plant; and then calculate the current values of certain evaluation criteria, such as fuel consumption, costs, and emissions. Later, after completing all the steps of synthesis, the program shows the changes in the values of those criteria obtained by optimization, and it prescribes the necessary changes to the use and loading of the plant.

A static calculation also provides a way of testing a computer program in the field.

Validate

The original research on the unified method included proof and examination of its validity for both the design and operation of energy systems (Tostevin and Luxton, 1979). Later, in industry, static calculations, followed by optimizations, continued to demonstrate the validity of the method.

Static calculations test both the computer program and the given information. Under steady conditions, such calculations should produce values that correspond with those showing on a plant's instrument and control system. Otherwise, it is necessary to reconcile any difference and to adjust either the computer program or the instrument system or both. Such reconciliation, conducted over a range of demands and plant availabilities, specifically validates the unified method for that one energy system. It also contributes to a general validation because the method is the same for all energy systems.

Validating the design of an energy system usually has to wait for construction, commissioning, and a period of operation. However, since the unified method is the same for both design and operation, experience gained from applications to the operation of existing energy systems provides the necessary confidence for design.

CHAPTER 6

Demonstrate

The Use and Benefits of a Unified Method for Operating an Industrial Power Station, Designing the Energy System for an Office Building, and Other Applications

These demonstrations are examples of practical applications of the unified method. They show that as far as the use of energy, plant, and their costs are concerned, we can

- Express the scientific, technical, and economic parameters of an energy system in universal terms, then
- Reliably conduct its engineering and evaluation in those terms, and consequently,
- Determine the best concept, design, and mode of operation for an energy system in any given situation

This is the essence of the unified approach. It is not necessarily comparable with other computer methods for energy systems because it synthesizes a range of different systems rather than model a particular system. It deals with the use of energy and plant over periods ranging from half an hour to a lifetime. It does not need to simulate transient or abnormal effects but expects an energy system to be built and equipped to manage those effects.

The demonstrations include two interactive computer programs, one for operation of an energy system and one for design, each with its database files, utility programs, and written instructions (see Appendix D). Both demonstrations derive from experience with the unified method in industry, but they do not represent any particular application. All the numerical and financial information is arbitrary, sufficient to demonstrate the method without necessarily applying to real practice.

Operate

Demonstration "Operate" is an example of the use of the unified method to optimize the operation of an industrial power station. It shows that the method can deal with the intimate details of an energy system and decide from options and alternatives that involve just the use and loading of individual sections and units of plant.

This demonstration requires a machinery database file but neither a time-base nor a tariff file because it is concerned with just the current operation of an energy system. It also employs a simple method for optimization: For a chosen initial objective, the processing records the detail of the progressive best solution, which, after completing all the steps of synthesis, defines the optimal solution.

Imagine an industrial power station, fueled by natural gas, meeting demands from a manufacturing plant for electricity and both high- (HP) and low-pressure (LP) process steam (Figure 6.1).

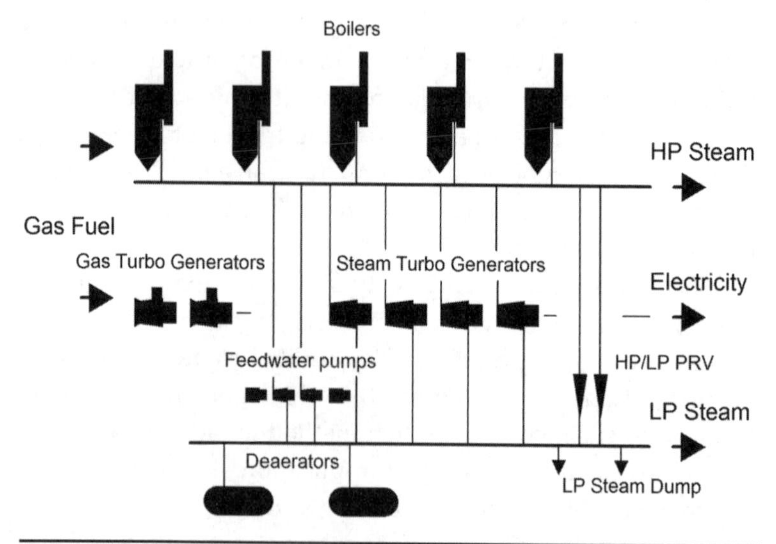

FIGURE 6.1 Industrial power station: configuration of plant.

Five boilers produce HP steam. Four backpressure steam turbo generators and two gas turbo generators produce electricity. Two HP/LP pressure-reducing valves provide additional LP steam if required, whereas dump valves can discharge surplus LP steam to the environment if necessary. Two motor and two HP steam pumps supply boiler feed water, taken from two deaerators. All this constitutes the maximum configuration of the plant within which some lesser, current configuration may operate at any one time.

A sketch of the cellular view of this energy system shows the transfers of energy from source to demands connected through cells representing each section of plant (Figure 6.2). The sketch also shows the values of numerical indices assigned to the source of energy, demands, and each section of plant. The sequence of transfers of energy directs the *order of calculation* for the computer program (see "Calculation" in Appendix C).

For a given objective, such as minimum fuel consumption, the demonstration optimizes the operation of the power station for either a set of given demands or a set of current loads on its plant.

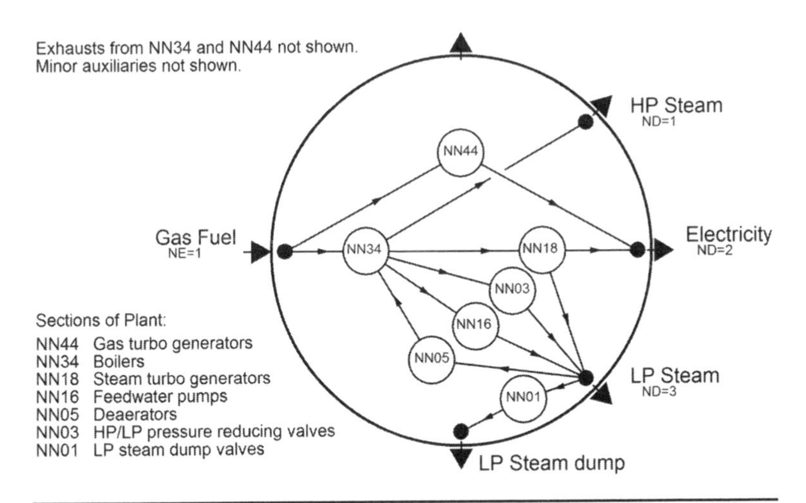

FIGURE 6.2 Industrial power station: cellular view.

Demands

Select the objective for optimization, and enter the availabilities of plant, other given and current information, and the values of demands. The computer program then synthesizes all the different possible ways to operate the power station and reports the best. Some output screens show the results for the whole power station; others show the result for each section of plant, especially the optimal use and loading of each unit.

Each use of the computer program reports just the one optimal solution for the given demands. However, the program can produce such solutions over a whole range of values for demands, plant availabilities, requested reserves, and conditions in the environment. A plot of such solutions shows profiles of minimum fuel consumption for the power station (Figure 6.3).

The profiles are not smooth; they reflect some of the practicalities of operating a power station in which a unit of plant is either working, at whatever load, or shut. Figure 6.3 also explains the necessary changes in the use of the plant as demands increase.

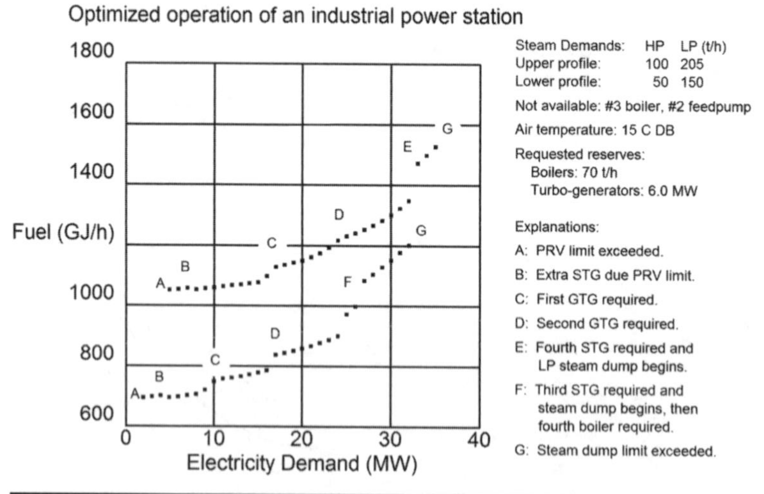

Figure 6.3 Power station: profiles of minimum fuel consumption.

There is little scope to optimize the operation of an energy system when its demands are high, with all the available units of plant near full load, because there is then virtually only one way to operate. As demands decrease or extra units become available, the scope for optimization increases.

Computing from demands is appropriate for scheduling the future operations of a power station. To optimize the current operation of a station, it is visibly more effective to compute from loads, selected by an entry to the first interactive screen.

Loads

After selecting an objective and entering given and current information, a loads screen then appears in place of the former demands screen. The loads screen requires manual entry of the current load on each unit of plant, although, in practice, a computer program might acquire the loads from an instrument and control system.

For the given loads, the processing first makes a static calculation of the current energy system and calculates the apparent demands. It follows immediately with an optimization for those demands and then compares current information with optimized information. Some output screens report the expected reduction in fuel consumption, emissions, and costs for the whole power station (Figure 6.4). Others report the current load on each section and unit of plant together with the changes required to achieve optimal operation.

If the current use and loading of the plant are already optimal, the processing reports that no further improvement is possible.

Design

Demonstration "Design" is an example of the use of the unified method to design the optimal energy system for an office building. It shows that the method can deal with a variety of

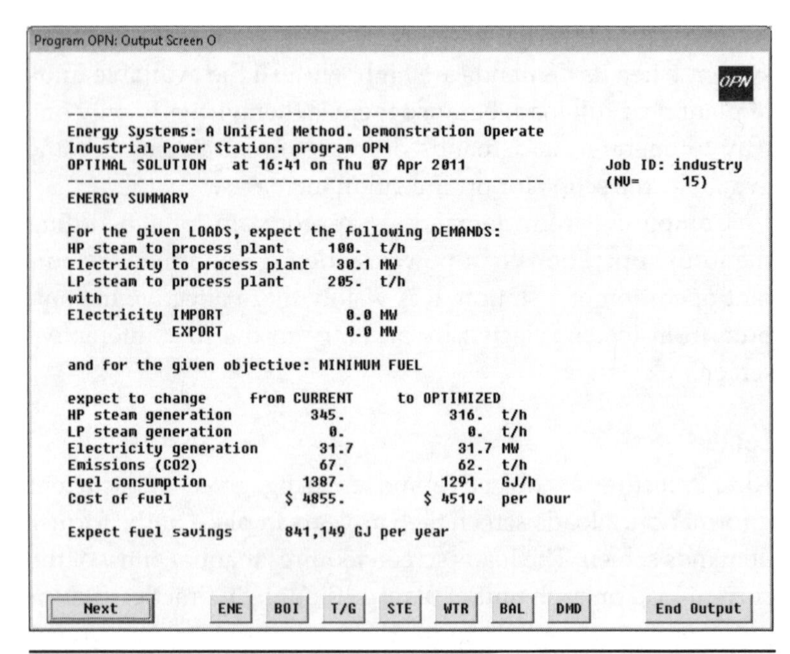

Figure 6.4 An output screen from the demonstration "Operate."

different kinds of plant and decide from options and alternatives that represent different concepts for an energy system.

This demonstration requires machinery, time-base, and tariff files. At each completed step of synthesis, the computer program records a design and its values of several different evaluation criteria. On completion of all the steps the program quickly finds and resynthesizes any one design or the optimal design for any chosen criterion.

Imagine a new multistory office building of about 30,000 m^2 floor area in a southern temperate climate for daytime occupation during the week and limited use on weekends. Its energy system has to meet demands for electricity (maximum 1,350 kW) and both hot and chilled water for air conditioning (maxima 5,500 and 9,500 MJ/h, respectively). Public supplies of gas and electricity are available (Figure 6.5).

The owners of such a building at least would expect to install gas-fired water heaters, electric motor–driven water

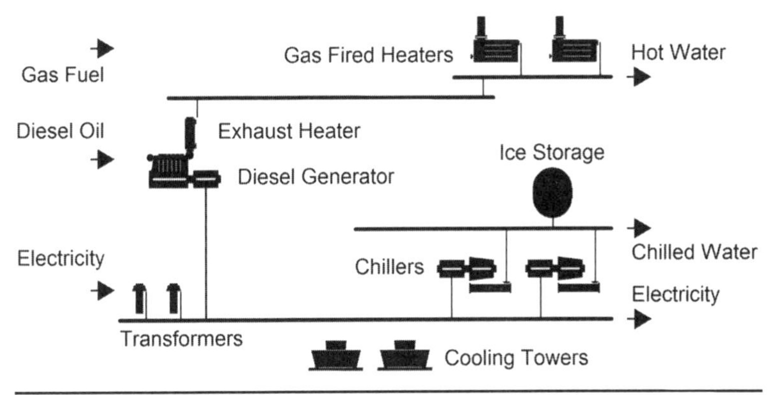

FIGURE 6.5 Building: maximum configuration of plant.

chillers, and cooling towers. However, they also may consider installing additional plant if it significantly reduces the lifecycle cost of the energy system. For example, ice storage would reduce both the rating and the peak electricity consumption of the chillers and use a night electricity tariff. A diesel generator of modest rating, using oil fuel, could meet part of the electricity demand in the event of failure of the public supply and even run continuously if required. An exhaust water heater on the diesel generator would reduce the rating and fuel consumption of the gas-fired water heaters. Altogether, these requirements and options prescribe the maximum configuration of plant for design.

This demonstration uses generalized information for the plant and minimum lifecycle cost as an initial objective for optimization.

After entering values for financial parameters and limits, the program synthesizes 90 different designs (Figure 6.6).

The program then sorts and ranks those designs according to several evaluation criteria (Figure 6.7). On request, the program resynthesizes any one design.

Design 3, with the highest cost of electricity, is the simplest energy system with just the minimum configuration of plant. All electricity is imported. Gas-fired heaters meet all the

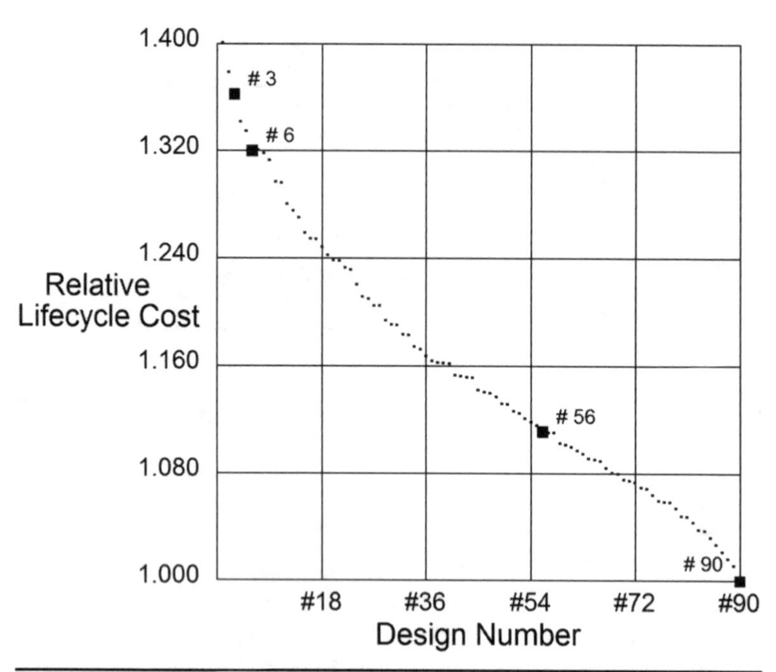

FIGURE 6.6 Building: 90 different designs.

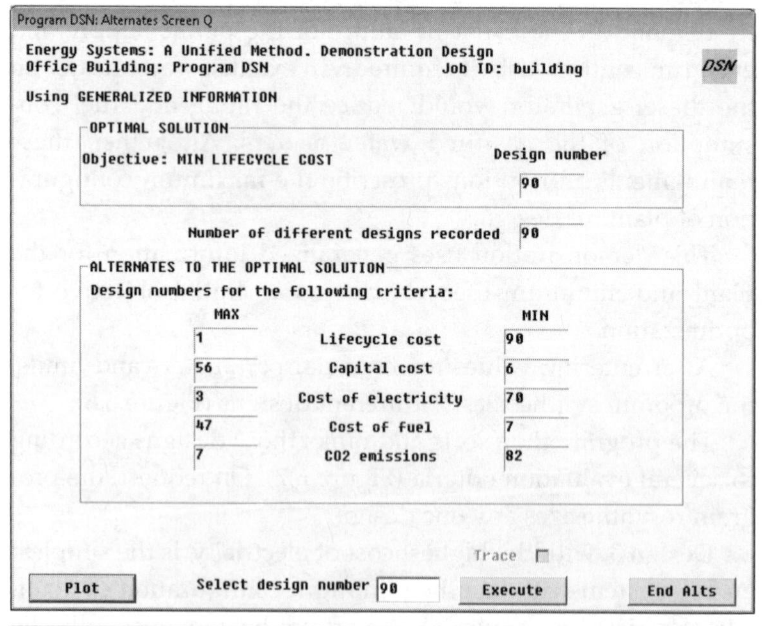

FIGURE 6.7 Alternates output screen from demonstration "Design."

demands for hot water. Electric chillers meet the demands for chilled water directly, without ice storage.

Design 6, of least capital cost, is similar to design 3 but includes 40 kL of ice storage, which reduces the rating of the chillers by 30 percent. The overall effect is to reduce the capital cost of the energy system by 8 percent and its lifecycle cost by 3 percent.

Design 56, of highest capital cost, includes all sections of plant in the maximum configuration, but in a relatively unbalanced way. A diesel generator helps to meet electricity demands during the day. An engine exhaust heater reduces the consumption of gas, but without reducing the ratings of the gas-fired heaters. Ice storage is used to obtain the benefits of a night electricity tariff without reducing the total rating of the chillers.

Design 90, of least lifecycle cost, also includes all sections of plant in the maximum configuration, but in a more balanced way than design 56. Ratings of the gas-fired heaters are reduced by 30 percent. Daytime use of ice storage reduces the ratings of the chillers by 30 percent.

Compared with the simplest design (design 3), the capital cost of design 90 is 4 percent less, and its lifecycle cost is 27 percent less.

All the preceding values for ratings and costs of the plant are arbitrary. They depend on the given information in the database files and the increments of size or rating applied during synthesis. They appear here merely as examples. Each energy project requires its own set of given information.

Each of the 90 different designs would work and meet the demands satisfactorily. However, their wide range shows how easy it might be to choose a poor design without a view of all possible designs, and it shows the benefit of such a view when faced with the need to make decisions about any one design.

For demonstration "Design," if lifecycle cost were the sole criterion, design 90 would be the best. However, decisions about new energy systems are not always so straightforward, and usually everyone concerned has to agree on a particular mix of evaluation criteria to determine the best system. It is a simple matter then to tailor the executive part of the computer program to define such an objective, for which the program then will find and resynthesize the optimal solution.

Other Applications

Although not demonstrated here, other applications tell of different ways of using the unified method.

Manufacturing

An industrial power station is usually only a part of a large manufacturing facility. However, we can apply the unified method to deal with the whole facility so that it minimizes the cost of energy for manufactured products rather than just electricity and steam.

We simply expand the boundary of the energy system to surround the whole facility. We identify each manufacturing process that significantly uses or produces transfers of energy. Then we convert their information into universal terms and include their plant in the maximum configuration. The required rates of manufacturing become the demands on the energy system.

Machines used for manufacturing often have irregular performance curves, for which we need numerical tables to provide values for their efficiency. Conversion factors are also unusual because they relate transfers of energy to flows of manufactured products. Otherwise, assembling information in the machinery database file is similar to that required for a power station on its own. Values of indices NN and NU continue to identify particular sections and units of the manufacturing plant, respectively. The time-base file may have to include data peculiar to the manufacturing processes. The functional specification for the computer program remains unchanged.

Private/Public Electricity

A manufacturing corporation often operates several process plants, each with its own industrial power station, connected to the public electricity grid. Contractual arrangements may allow the corporation to export and import electricity to and from the grid at certain times to help maximize the use of its plant and minimize overall costs.

Again, we expand the boundary of the energy system to surround all those power stations, and the unified method deals with them as a group. For each power station, the method first minimizes waste and internal costs. Then it determines the minimum need to import electricity and the maximum capacity to export. It then makes the most of opportunities to "wheel" electricity from power stations with excess generating capacity to others with a deficiency.

The extent of these opportunities varies over time, as demands on the power stations change with rates of manufacturing, as availabilities of the plant change with maintenance requirements, and as limits, charges, and credits for import, wheeling, and export of electricity change by the hour of the day and the day of the week. For all this, the unified method continually minimizes the net total cost of electricity and fuel for all the corporation's plants combined.

CHAPTER 7

Critique

Simulation and Optimization, Questions and Answers

Initiating and developing the unified method taught many lessons. At first, it seemed straightforward: Top academic examiners approved the original research, and industrialists appreciated the cost savings obtained in the operation of their plants. The more complicated the plant, the more the appreciation.

For new energy systems, people valued the presentation of a variety of different designs from which they could choose the best. However, they rather expected consulting engineers to do this anyway, whatever their method. They particularly welcomed designs that included energy storage, for which a comparable method seemed not to exist.

Yet many people doing the detailed work on energy systems did not so readily accept the unified method. This seemed to be due in part to a natural reticence to change their current way of working. It also possibly reflected a training that understandably leaned toward methods of analysis rather than synthesis. It became necessary to emphasize that the unified method was doing things that otherwise would not be done and that the effort required for a first project would be virtually automated for future projects. Acceptance grew as people became familiar with the method and as the power of their computers increased. This chapter deals with some of the questions that arose.

Simulation

People usually think of simulating energy systems with mathematical models. In contrast, the unified method simulates energy systems with assemblies of cells. All the cells are similar, each with the same set of containers, although with different numerical content. Every assembly of cells has a single common thread: its rates of transfer of energy. An assembly can change from moment to moment owing to variations in those rates, or in the content of containers, or in response to programmed logical instructions.

Such a simulation is rather like the way a computer itself works, in terms of words, their numerical content, and logic. This becomes evident in practice, when we see a computer program for the unified method dealing directly, and in a similar way, with all parts and conditions of an energy system.

A computer program for the unified method still incorporates mathematical expressions when they are necessary. The demonstrations in this book use them to represent the performance of units of plant and to provide values for steam tables. It is also likely that some people will prefer in the future to conduct certain parts of the unified method mathematically, the details of which will evolve.

Optimization

The design of energy systems has been a discipline of mechanical engineering for some time. Traditionally, its method applies optimization procedures to mathematical models of energy systems (Stoecker, 1989).

Questions frequently arise about the merits of different methods. Application of the traditional method is rather individual for each energy project, whereas the unified method is virtually the same for all projects. The former deals with just one predetermined concept for an energy system at a time and finds its way mathematically to the optimal design for that concept (Stoecker, 1989, p. 144). In contrast, the unified method synthesizes all possible designs for all concepts of energy systems within the scope of its given information. It identifies the best for any subsequently chosen objective, and it displays the detail of any one design on request. This ability to determine the best concept for an energy system, with virtually global scope, is the vital difference and promise of the unified method. It also ensures that no opportunity for design or operation is overlooked, and it presents solutions that otherwise may not be conceived.

Yet the two methods are not rivals; each has its place. For example, once the unified method determines the best concept for an energy system, people still may use the traditional method for its physical engineering.

Questions also arise about the way the unified method seems to use just a single objective for optimization. In one sense, it does, because it consistently minimizes the use of both energy and plant in every unit time. The former seems to be logical for any procedure concerned with the use of energy, and the other is resolved by imposing a different plant at each step of synthesis. In a wider sense, however, the unified method does not apply its optimization until all the steps of synthesis are completed. Definitions of the objective then may take virtually any required form and, indeed, be prompted by the results of the synthesis itself.

Computing

In the 1940s, it was not so apparent how the view of state-couples in the processes of energy systems could be useful, but that was before the general availability of computers. Later, during the original research for the unified method in 1979, the set of different designs for a modest energy system required an overnight run on a mainframe computer. Since then, the power and speed of computers have increased steadily so that, in 2012, the same set of designs takes only a few seconds on a notebook computer.

Meanwhile, the unified method has not changed in principle. Yet the detail of its computer program has changed owing to experience gained from applications in practice and the evolution of methods for personal computing.

"Where is the computer software?" has been a common question arising from applications of the unified method in industry. The immediate answer is that it can be downloaded from the website at www.mhprofessional.com/TostevinEnergySystems.

However, the principal question remains—about the increase in computing power needed to deal with energy systems on, say, a national scale. Such an increase in the power of personal computers indeed may occur, as it has in the past. On the other hand, computing for the unified method is a vast but highly organized repetition of similar, almost trivial numerical calculations. For the demonstrations in this book, the computer program conducts those calculations sequentially, but it could just as well conduct them in simultaneous parallel streams and collate their results on completion. Computer programs for applications of the unified method on a wide scale then would share their work among multiple processors. This seems to present an opportunity for future research and development.

For the time being, the unified method can still use an overnight run on a computer to synthesize, evaluate, sort, and record a vast number of different unions, designs, or modes of operation of energy systems. Next morning, the computer presents a complete field of results from which to select and resynthesize for any given criteria.

Some people ask about the fundamental limit of the unified method. No matter how high the computing power, the method cannot synthesize, in a finite time, all possible energy systems within a region of unlimited scope. However, it has a practical way of dealing with this. It allows us to delineate a *region of interest* for each application by defining both a maximum configuration of plant and its increments of change. Within this region, a computer readily can synthesize all possible energy systems. Other things being equal, the region of interest can expand as computing power increases.

Application

This book frequently has remarked that the unified method is the same for all energy systems, at least those of the physical sciences. People justifiably question this, especially when they cannot find an example of its application to an energy system with which they are familiar.

Part of the answer is that we deliberately based the original research of the unified method on what seemed to be an energy system of the worst case: a solar and fuel-fired water-heating system combining a time-varying source of energy with a time-varying demand and using energy storage. Since then, applications to a variety of industrial and commercial energy systems have not presented a worst case, except in their extent, diversity, and detail.

Another part of the answer is that every application seems to introduce at least some kinds of processes and plant that the unified method has not met previously. Converting their information into universal terms, bringing them into the unified method, and seeing the result is a rewarding experience.

APPENDIX A

Algorithm

All parts of a computer program for the unified method have to work together quickly, reliably, and repeatedly in each unit time through the period of integration for each step of synthesis. An algorithm built into the structure of the computer program regulates this (Figure A.1).

The following notes summarize how each principal part of the computer program fulfills the requirements for that algorithm.

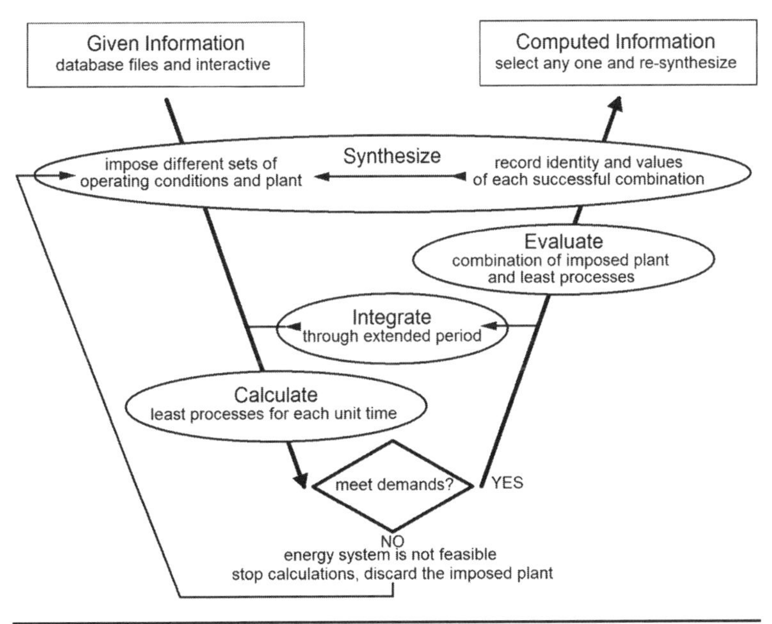

FIGURE A.1 Algorithm of computer program.

Synthesize

- Impose certain operating conditions for the energy system, and if necessary, vary their values incrementally through a practical range.
- For each set of operating conditions, also impose different practical configurations of the plant by applying incremental reductions to the size or rating of each section of plant in turn.
- Each different set of imposed conditions and plant defines a step of synthesis, identified by the value of its index, symbolic name NV.

The processing imposes different sets of values for operating conditions by assigning values for independent temperatures and pressures in containers TT and PP. The processing imposes different configurations of plant by assigning values in container SIZE for each section of plant in the maximum configuration. It first imposes a certain high value of physical size, rating, or number of units and then reduces it by increments down to a certain minimum or zero. It applies such reductions to all sections of plant in turn so that it imposes all practical configurations of the plant. This enforces evaluation of the effects of different ratings of plant. It also enforces evaluation of alternative sequences of processes when different sections of plant are able to share a needed transfer of energy.

Imposing different sets of values for operating conditions and plant is the way the processing systematically changes from one energy system to another. It might impose thousands of different sets of values, but the calculations are simple, similar, and repetitive—and many are quickly discarded.

Usually, it is neither necessary nor appropriate to synthesize the design of all possible energy systems for any one purpose. Instead, we control the scope of the processing with the notion of a *maximum configuration* that admits just the options and alternatives—sources of energy and kinds of plant—that people are prepared to consider for each project. In essence, the maximum configuration represents the scope of the given information.

Calculate

- For each step of synthesis and for each unit time, calculate the rates of transfer of energy of the least processes needed to meet the demands and, finally, their supplies from sources of energy. For design, and if necessary, conduct such calculations for both maximum and average demands concurrently.
- Discard the step of synthesis and its imposed configuration of plant as soon as it becomes apparent that the latter is unable to contain the necessary processes and therefore cannot meet the demands.

For each set of imposed conditions and plant, for each unit time, and beginning with the values for demands, the processing works upstream through the cellular view of the energy system toward the sources of energy in a logical order. It calculates the least rate of supplied transfer of energy at each cell in turn, which then becomes the rate of needed transfer of energy from the next cell upstream. It reiterates the whole or parts of the calculation if some processes affect those previously calculated (see "Calculation" in Appendix C).

For each step of synthesis, the processing attempts but does not necessarily complete such calculations. Eventually, as the imposed ratings of the plant reduce, a particular cell will be unable to contain the required processes, and the continuity of transfers of energy fails. The energy system is not feasible. The calculations then stop, and the processing discards the imposed plant configuration. Otherwise, the processing calculates the rates of supply from the sources of energy. The calculated energy system will be the one that is just able to meet the given demands and no more, and the consumption of energy will be the least possible for the imposed conditions and plant.

The results of completed calculations, tested with an energy balance, describe the details of the processes in the imposed plant for the unit time.

If a cell represents just a single unit of plant, calculating the least rate of supplied transfer of energy is straightforward because there is just one solution for the values in the cell's containers. For a cell representing a section of plant with multiple units, however, the processing must distribute the needed transfer of energy optimally among the available units, even if it has to exclude (or shut) some of them (see "Machine Loading Subroutine" in Appendix B).

Integrate

- Progressively integrate the results of successful calculations for each unit time through whatever extended period is required for evaluating an energy system or managing its operations.
- If energy storage is included in the maximum configuration of plant, conduct the integration in reverse order of time.

For each imposed configuration, the processing continues the integration only as long as the calculations for each unit time are being completed. When the calculations stop, the energy system is not feasible, and the processing discards the results of integration along with the imposed configuration.

Integration is always required when processing through an extended period, and operations with energy storage always require integration through each store cycle. The period of integration also may depend on the requirements for evaluation; costs of fuel, for example, usually require monthly integrations, whereas lifecycle costs need yearly integrations.

The results of a completed integration, tested with another energy balance, describe the operation of the imposed configuration of plant through the extended period.

With *energy storage* in the maximum configuration, and integrating in reverse order of time, the required size of storage evolves from the reductions of imposed sizes or ratings of plant as follows:

- Eventually, at some unit time, and at the point of storage, the processes in the section of plant of imposed size or rating cannot meet the rate of needed transfer of energy. Instead of allowing the continuity of transfers of energy to fail, draw the deficiency from storage.
- Then, for the succeeding unit time, include in the calculations for the energy system a supply to storage of whatever replenishment is required or available from a connected section of plant upstream.
- Integrate through the store cycle the net effects on storage of the transfers of energy drawn, supplied, lost, and gained. This determines the current value of *energy-in-store* at any one unit time. During integration, that value rises to a maximum, which represents the size of storage necessary for the energy system to remain feasible.
- Index the integration twice through the store cycle, and verify on completion that the value of energy-in-store is the same as that at the beginning. If not, the energy system is not feasible. Further, at any time during the integration, if the required size of storage exceeds any given limit, the energy system is not feasible.

The size of storage evolved from the preceding will be just the minimum needed to keep the energy system feasible; this is essential or "active" storage. However, the conditions of supply for some sources of energy may justify even more storage, which is merely helpful or "passive" storage. When the cost of a source of energy varies with the time of day, for example, an energy system may have the capacity to work and store the effect of an excess supply at some times to reduce costs at other times. For this, the processing has to evaluate the energy system both with and without such an extra use of storage.

Evaluate

- On completion of the integration, evaluate for several different criteria the combination of imposed plant and least processes, each of which represents a different energy system in terms of either its design or mode of operation or both.
- Discard an energy system if its costs, emissions, or other consequences exceed any given limits.
- Otherwise, record the identity NV of each successful energy system together with its values for evaluation criteria.

Each successful step of synthesis produces a complete description of the design and operation of a particular energy system; and the processing calculates and presents the values of evaluation criteria for each of them. The wider the range of objectives of the people who use the computer program, the wider the required variety of criteria.

The processing calculates the costs of sources of energy from their consumptions and prevailing commercial tariffs; if alternative tariffs are available, it adopts the one that yields the least total cost. It calculates emissions from the consumptions of fuels and their chemistry of combustion. It calculates the costs, space, and weight of units of plant from information in containers VV.

The processing combines component costs, in whatever required form, to determine total values for the whole energy system. For design, this includes capital, annual fixed and annual running costs for the plant, and annual costs for supplies of sources of energy. For operating an existing energy system, just the current and monthly costs of sources of energy may be required.

The processing discards a design or mode of operation if evaluation shows that its costs or other consequences exceed any given limits.

Optimize

- On completion of all the steps of synthesis, rank all the successful energy systems according to their values of evaluation criteria, and identify the best and worst for each of them.
- On request, select and resynthesize any one energy system and display its detail: operating conditions, consumptions of sources of energy, emissions, sections and units of plant, loads, annual utilizations, space, weight, and all costs.
- The detail of the energy system that best meets a given objective, defined by one or more evaluation criteria, constitutes the optimal solution.

APPENDIX B

Specifications

This appendix specifies the detailed functions required of a computer program for the unified method, all in accordance with the algorithm of Appendix A.

To the extent specified here, and subject to the evolution of new techniques in the future, these functions are virtually the same for all applications of the unified method. Use the specification both as a guide to understanding the source code of the example computer programs and as a framework for building new programs.

Each principal part of the computer program shown in Figure 5.2 is a separate subprogram. The following pages list the functions required for each part, beginning with the executive and working toward the calculation subprogram. The specification for each part begins on a new page, and illustrates its structure in the C++ programming language.

The *executive* starts, stops, and manages the computer program. It reads given information from the database files, and it conducts an interactive dialog to receive additional information. It writes computed information to output files and interactive screens.

The executive directs the processing required for optimization, particularly to select and resynthesize the energy system that best meets certain criteria.

The executive also may invite interactive entry of an initial objective for optimization, for which the executive continuously records the details of the progressive best solution. After completing all the steps of synthesis, the progressive best solution then describes the optimal solution for that objective. This feature also monitors the health of the computer program during synthesis and provides a starting point for examination in the event of a problem.

The *synthesis subprogram* imposes different sets of operating conditions and configurations of the plant in orderly steps. It dictates rather than calculates values for the computer program. For each step of synthesis, the processing produces either a solution, for a feasible energy system, or no solution. It records the identity of each solution in an output file, along with its values for evaluation criteria. After completing all the steps of synthesis, the subprogram initiates a diagnosis if the processing has not recorded any solution.

For each step of synthesis, the *evaluation subprogram* calculates the values of several criteria that record the costs and other attributes and consequences of each solution. It normally returns those values to the synthesis subprogram to become part of the detail of a solution. However, if those values exceed any given limits, it returns "No solution" to the synthesis subprogram.

For each step of synthesis, the *integration subprogram* directs the series of calculations required for each unit time through an extended period, and it accumulates the results. On satisfactory completion of all such calculations, it returns integrated values to the synthesis subprogram. However, if calculations fail for any unit time, the integration stops, discards results, and returns "No solution" to the synthesis subprogram.

For each unit time, the *calculation subprogram* computes the rates of transfers of energy through the energy system, and it normally returns the details of that energy system to the integration subprogram. However, if the continuity of transfers of energy fails, the energy system is not feasible, the calculation stops and returns "No solution" to the integration subprogram.

Executive

For the demonstrations of this book, the executive combines an interactive dialog in C++ with a procedure in FORTRAN to conduct the following functions:

```
main()
{
    // Part A: Given information.

    // Part B: Synthesize.
    // Call the synthesis subprogram.
    synthesize();
    // Either receive the identity of the
    // progressive best solution,
    // or receive a diagnosis if no solution.

    // Part C: Output information.
    // Display and record the details of the
    // progressive best solution
    // or the result of the diagnosis.
    // Sort and analyze all solutions.

    // Part D: Optimal and alternate solutions.
    // Select any one solution that best meets
    // certain evaluation criteria, and recall
    // synthesis subprogram to obtain its detail.
    // Display and record the selected solution.
}
```

Part A: Given information.

- Start the program, initialize variables.
- Open one or more output files for recording results.
- Open, read, and close the database files.
- Test that the given information is complete; if not, display messages and request more information.
- Write a summary of the given information to the output file.

Interactive.

Select an initial objective for optimization if required; then

Executive (*continued*)

Either, for operation,

- Select calculation from either given ***demands*** or current ***loads***.
- Enter the current conditions in the environment and any current limits or restrictions to the sources of energy or demands.
- Enter any financial values that currently apply.
- Enter the current availabilities of units in each section of plant.
- Enter any other current information and requests for reserve capacities, such as spinning reserve for turbo generators.
- Either enter the current demands or, if computing from loads, enter the current load on each unit of plant.

Or, for design,

- Select the information to be used from the machinery database file, either generalized or specific.
- Select the extended period of operation of the energy system.
- Select sources of energy, demands, and sections of plant and so define the maximum configuration.
- Enter financial values required for the evaluation of costs.
- Enter adjustments to rates of demand, costs of sources of energy, and costs of plant, if required for sensitivity tests and to answer "what if" questions.
- Enter any restrictions that have to be met by the design, such as maximum or minimum numbers of units to use in a section of plant.

Executive (*continued*)

Interactive trace.

- If required, select a trace of the execution of the program that displays data and messages at each step.

Part B: Synthesize.

- Call the synthesis subprogram.

On completion of all the steps of synthesis:

- **Either** receive the identity and details of the progressive best solution, including its values for evaluation criteria,
- **Or** receive a diagnosis of why no solution has been obtained.

Part C: Output information.

- Display and write to the output file the identity and details of the progressive best solution or, in the absence of a solution, the result of the diagnosis.

Interactive.

- Display the details of the progressive best solution, including
  - Its configuration of plant, sources of energy, demands, transfers of energy, values for evaluation criteria, and costs, together with
  - A table or diagram for each section of plant showing its transfers of energy and costs; and
  - For *operation*, the availability, current load, and optimal load of each unit of plant, together with a diagram showing the rates of flow of the principal working fluids through the whole plant, such as the distribution of steam.

Executive (*continued*)

- For *design*, the hours of use and annual utilization of each unit of plant.

- Sort and analyze all solutions, and identify, for example, the best and worst for several evaluation criteria.

Part D: Optimal and alternate solutions.

- Select any one solution for examination in detail, including the best or optimal solution for given criteria, and call the synthesis subprogram to resynthesize its energy system.
- Display and write to the output file the details of the selected solution as listed earlier for the progressive best solution.

END of executive.

- **Either**, repeat the whole program from Part A,
- **Or**, close the output file and *stop the program.*

Synthesis Subprogram

Called by the executive to conduct the following functions:

```
SYNTHESIZE()
{
  int PF, PFMAX, PFMIN; // See Dictionary.
  int NIMP, NAVL; // See Dictionary.
  // Part A: Impose values in containers.
  // Operating conditions: values in PP, for example.
  for ( PF = PFMAX; PF >= PFMIN; PF-- )
  {
    // Impose configurations of plant.
    // Limits of value in SIZE
    // as numbers of units of plant, for example.
    for ( NIMP = NAVL; NIMP >= 0; NIMP-- )
    {
      // Steps of synthesis.
      // Generalized or specific information.
      // Part B: Integrate.
      // Call the integration subprogram.
      integrate();
      {
        // Part C: Evaluate.
        // Call the evaluation subprogram.
        evaluate();
        // Solution is within given limits.
        {
          // Record progressive best solution.
          // Record each solution
          // in an 'alternates' output file.
        }
      }
    }
  }
  // Part D: No solution has been obtained.
  // Restart the indexing from Part A
  // to obtain a diagnosis.
  return;
}
```

Part A: Impose values in containers.

Impose operating conditions: Values in TT and PP, for example.

- **Either**, fix independent temperatures and pressures of processes at standard conditions,
- **Or**, index incremental changes over a range of values.

Synthesis Subprogram (*continued*)

For example, evaluate the working of an energy system at three alternative fluid pressures. Index such imposed conditions as an outer iterative loop to those below.

Impose configurations of plant: Limits of value in SIZE, as numbers of units of plant; for example:

- Index all sections of plant in the maximum configuration, arranged as a nest of iterative loops, one for each section of plant, matched to the order of the calculation subprogram.
- **Either**, index the number of units imposed, beginning with the maximum available and ending with a certain practical minimum. For example, if seven units are available, first impose seven, then six, and so on down to two or one or zero.
- **Or**, for design, if the number of units in a section of plant is fixed, index reductions to the limits of value in SIZE by physical increments.

Energy storage.

- If energy storage is included in the maximum configuration, index so that it is alternately included with and excluded from each of the preceding sets of imposed conditions and plant.

Steps of synthesis.

- Identify each unique set of imposed conditions and plant as a step of synthesis; index NV.

Generalized or specific information.

- For operating an existing energy system, assign specific information.

Synthesis Subprogram (*continued*)

- For design, automatically allocate ratings and numbers of units required for each section of plant by reference both to the maximum demands and to any interactive requests. Initially assign generalized information to all such sections of plant, but replace it with specific information if the latter is requested (see "Generalized and Specific Information" in Appendix C).

Bounds on synthesis.

- Apply bounds to limit the scope of synthesis, if required, but include a switch to test results with and without such bounds.

Part B: Integrate.

- Call the integration subprogram.

For each set of imposed conditions and plant, either the energy system remains feasible through the period of integration or it does not.

Energy system is* not feasible*:

- Exit from the innermost of the preceding iterative loops, and continue the indexing of Part A.

Energy system is* feasible*:

- Obtain the details of the integration.

Part C: Evaluate.

- Call the evaluation subprogram.

For each evaluation, either the energy system meets any given limits of value, such as maximum capital costs, or it does not.

Synthesis Subprogram (*continued*)

Solution is not within given limits:

- Exit from the innermost of the preceding iterative loops, and continue the indexing of Part A.

Solution is within given limits:

- Record the identity and detail of the progressive best solution.
- Record, in an alternates output file, the identity of each solution, index NV, and its values for evaluation criteria.

Part D: After completing the indexing of Part A, no solution has been obtained.

- Set an indicator that enables the generation of diagnostic messages throughout the whole computer program, and restart the indexing of Part A.
- Stop the indexing of Part A on receipt of the diagnosis.

END of synthesis subprogram.

- Return to the executive either the identity and detail of the progressive best solution or an explanation from the diagnosis if no solution has been obtained.

Evaluation Subprogram

Called by the synthesis subprogram to conduct the following functions for each completed step of synthesis:

```
EVALUATE()
{
    // Part A: Evaluation criteria.
    // Costs of sources of energy.
    // Costs of plant.
    // Costs of whole energy system.
    // Emissions.
    // Part B: Other objectives.
    // Values for other criteria.
    // Part C: Limits on solutions.
    {
        // Limits exceeded.
        // No solution.
        return;
    }
    return;
}
```

Part A: Evaluation criteria.

Costs of sources of energy.

- Apply either specific conditions of contract or each of any available alternative tariffs to the supply from each source of energy through the extended period. For each source of energy, use and identify the contract or tariff that yields the least total cost, and record its identity.

Costs of plant.

- For each section of plant and each of its units, apply the values in container VV according to whether the unit is used at all through the extended period and, if so, to its hours of use.
- Add the components of cost for all units in each section of plant.

Evaluation Subprogram (*continued*)

Costs of whole energy system.

- Combine the costs of sources of energy and the plant according to the definition of any evaluation criterion that needs it, such as annual cost of fuel or lifecycle cost.

Emissions.

- Calculate the annual emissions of carbon dioxide and other gases and waste if required.

Part B: Other objectives.

- Calculate the values of any other evaluation criteria, such as the space and weight of the plant.

Part C: Limits on solutions.

Limits exceeded.

- If the value of any evaluation criterion, such as costs or emissions, or any other consequence of a design or mode of operation fails to meet any given limit, return "No solution" to the synthesis subprogram.

END of evaluation subprogram.

- Return to the synthesis subprogram the values of evaluation criteria for the current step of synthesis, index NV.

Integration Subprogram

Called by the synthesis subprogram to conduct the following functions through the extended period:

```
INTEGRATE()
{
  // Part A: Extended period.
  int IH, IE; // See Dictionary.
  for ( IH = 1; IH <= IE; IH++ )
  // Energy storage: reverse this indexing.
  {
    // Part B: Unit time.
    // Call the calculation subprogram.
    calculate();
    // Energy system is not feasible.
    {
      // No solution.
      return;
    }
    // Energy system is feasible.
    {
       // Part C: Progressively integrate.
       // Transfers of energy.
       // Hours of operation.
       // Sources of energy and demands.
       // Test limits of value.
          // Limits exceeded: return no solution.
       // Record intermediate values.
       // Energy storage: see note.
    }
  }
  // Part D: End of extended period.
  // Annualize.
  // Energy balance for the extended period.
    // Balance fails: stop the program.
  return;
}
```

Part A: Extended period.

- Index each unit time through the whole extended period.

For a unit time of one hour, for instance, and for an extended period of one day, index from 0100 to 2400. For an extended period of a full year, index from 0100 to 2400 for each of 365 days.

Integration Subprogram (*continued*)

Energy storage.

- If active energy storage is included in the maximum configuration, *reverse* the preceding indexing so that it begins with the latest and ends with the earliest unit time of the store cycle.

For a unit time of one hour with an extended period and store cycle of one day, and with a datum time of, say, 2200, index from 2200 back through 0100 to 2300. Further, index *twice* in succession through each store cycle in which such storage is active.

Part B: Unit time.

- Call the calculation subprogram.
- For each unit time, either the energy system is feasible or it is not.

Energy system is* not feasible*:

- Stop the indexing of Part A and return "no solution" to the synthesis subprogram.

Energy system is* feasible*:

- Obtain the details of the energy system calculated for the unit time.

Part C: Progressively integrate.

Transfers of energy.

- Accumulate the values of the transfers of energy for each section and unit of plant.

Hours of operation.

- Accumulate the number of hours of operation for each unit of plant.

Integration Subprogram (*continued*)

Sources of energy and demands.

- Accumulate the values of supplies from each source of energy, and for deliveries to each demand.

Test limits of value.

- If the integrated supply required from a source of energy exceeds any given limit of value, the energy system is not feasible; stop the integration, and return "No solution" to the synthesis subprogram.

Record intermediate values.

- Record any intermediate values, such as hourly maxima and monthly totals, needed by the evaluation subprogram and tariff subroutines to determine the costs of sources of energy.

Energy storage.

- The calculation subprogram itself accumulates the transfers of energy to and from storage through each store cycle.

Part D: End of extended period.

Annualize.

- For design, annualize all integrated values, if necessary.

Energy balance.

- Test the results of integration by equating the values of transfers of energy entering and leaving each section of plant and the whole energy system for the extended period.
- If the error in balance exceeds a certain tolerance, display an explanatory message, and *stop the program.*

Integration Subprogram (*continued*)

END of integration subprogram.

- Return the details of the integration to the synthesis subprogram.

Calculation Subprogram

Called by the integration subprogram to conduct the following functions for each unit time:

```
CALCULATE()
{

  // Part A: For each demand ND
  // and its connected sections of plant in order of calculation.
  // Needed transfers of energy QN.
  // Share QN.
  // Part B: For each section of plant NN
  // in order of calculation.
  {
     // Section of plant subroutine.
     // Test limits of value.
       // Limits exceeded: return no solution.
     // Overload.
       // Unresolved: return no solution.
     // Load units of plant.
       // Unresolved: return no solution.
  }
  // Part C: Energy storage.
  // Accumulate through the store cycle.
     // Unresolved: return no solution.
  // Part D: Whole energy system.
  // Internal demands.
  // Transfers of energy.
  // Part E: Sources of energy, NE.
  // Limits of value.
     // Limits exceeded: return no solution.
  // Energy balance for the unit time.
     // Balance fails: stop the program.
  return;
}
```

First, define a logical *order of calculation* for the maximum configuration of the plant from demands ND through sections of plant NN to sources of energy NE (see "Calculation" in Appendix C).

For design, conduct calculations for both average and maximum demands concurrently.

Calculation Subprogram (*continued*)

Part A: For each demand ND and its connected sections of plant in order of calculation.

Needed transfers of energy, QN.

- Beginning with the rate of each demand QD[ND] and working upstream toward the sources of energy through the sequence of cells, determine the rate of needed transfer of energy QN for each section of plant in turn.
- For the section of plant that delivers to a demand, the value of QN is initially that of QD[ND]. For each section of plant upstream, the value of QN is initially that of QS previously calculated for the section of plant downstream.
- Initial values for QN may be recalculated later during reiteration.

Share QN.

- If two or more sections of plant are able to meet a needed transfer of energy, take a maximum share to one, pass the remainder to the next, and so on in the order of calculation until the total value of QN is fully accounted for. Subsequent systematic changes in the imposed plant will evaluate the sharing of QN by different sections of plant.
- In addition, and if necessary, reiterate calculations for two or more sections of plant if they require finer increments for sharing QN than are provided by changes in the imposed plant, such as, for example, to maintain at least a certain minimum load on a section of plant.

Calculation Subprogram (*continued*)

Part B: Each section of plant NN in order of calculation.

- Call its *section of plant subroutine* (described below) to assign values in containers for both the whole section and each of its available units.

Test limits of value.

- If any imposed or current value exceeds a given limit, the energy system is not feasible; stop the calculations, and return "No solution" to the integration subprogram.

Overload.

- Test that the total rating available from the number of imposed units of plant is enough to meet the value of the rate of needed transfer of energy QN. If not, the section of plant is potentially overloaded. However, if QN can be shared with another section of plant, take whatever maximum can be met.
- If the value of QN cannot be met or shared, and if energy storage is both included in the imposed configuration and available to the overloaded section of plant, take whatever maximum can be met or shared, and obtain the excess as a *draw* from storage. Note that this is active, not passive, storage.
- Otherwise, the energy system is not feasible; stop the calculations, and return "No Solution" to the integration subprogram.
- For a section of plant that can *replenish* storage, include in its calculations a supplied transfer of energy to storage for whatever replenishment is required and available.

Calculation Subprogram (*continued*)

Load units of plant.

- Distribute the value of the needed transfer of energy QN optimally among the imposed units of plant to minimize the rate of supplied transfer of energy QS (using the machine loading subroutine described below).
- Calculate dependent temperatures and pressures, and the rates of flow of working fluids.

Note that this ends the part of the calculation subprogram that deals with each section of plant in the order of calculation.

Part C: Energy storage.

- Progressively accumulate through the store cycle the effects of the *draws* and losses from storage and their *replenishment*.
- In each unit time, test that the size of storage remains within any given limits. At the end of the second indexing of each store cycle, test that the value of energy-in-store is the same as that at the beginning. If either test fails, the energy system is not feasible; stop the calculations, and return "No solution" to the synthesis subprogram (see "Energy Storage" in Appendix C).

Part D: Whole energy system.

Internal demands.

- Where necessary, reiterate calculations for one or more demands to account for any internal demands of the same kind (see "Calculation" in Appendix C).

Calculation Subprogram (*continued*)

Transfers of energy.

- Record the values of the rates of needed, supplied, residual, parasitic, and auxiliary transfers of energy for each unit and section of plant in turn, and hence for the whole energy system.

Part E: Sources of energy, NE.

- After completing the calculations for all demands ND, calculate the rate of supply required from each source of energy.

Limits of value.

- If the rate of supply from a source of energy exceeds any given limit, the energy system is not feasible; stop the calculations, and return "No solution" to the integration subprogram.

Energy balance.

- Test the results of calculations by equating the values of the transfers of energy entering and leaving each section of plant and the whole energy system for the unit time. If the error in balance exceeds a certain small tolerance, display an explanatory message, and *stop the program.*

END of calculation subprogram.

- Return the details of the calculation for the unit time to the integration subprogram.

Section of Plant Subroutines

Part B of the calculation subprogram calls these *subroutines*—one for each section of plant in turn—to conduct the following functions in support of its task headed "Each section of plant NN in order of calculation":

Containers.

- Assign values in containers for all available units in the section of plant, some from the machinery database file, others from connected sections of plant, some from conditions imposed by the synthesis subprogram, and others from the conditions of the environment.
- Where necessary, adjust the values from standard to current conditions by reference to mathematical expressions or numerical tables.

Load control.

- Identify the method of load control for units in the section of plant, such as proportional loading, working or shut, or incremental loading.

Reserve.

- Provided that the number of units imposed is less than the number available, apply any reserve capacity requested from the executive as a margin to the total rating of the imposed units.

Load units of plant.

- Call the *machine loading subroutine* (described below) to distribute the value of the needed transfer of energy QN optimally among the imposed units of plant to minimize the rate of supplied transfer of energy QS.
- Determine the rates of transfer of energy for the whole section of plant and all units.

The form of information in and assigned by these subprograms is similar for each section of plant. However, their details usually differ, so a similar subroutine is written for each section of plant, with names such as NN44() for gas turbo generators and NN18() for steam turbo generators. The source code for the demonstrations in this book provides examples.

Machine-Loading Subroutine

Called by section of plant subroutines to conduct the following functions in support of their task headed "Load units of plant." The source code of the machine loading subroutine is identical for all sections of plant in the demonstrations of this book.

Static calculation.

If requested by the executive.

- Fix the loads on units of plant at their current values, and return to the calculation subprogram the current values of the needed, supplied, residual, parasitic, and auxiliary transfers of energy for each unit and the whole section of plant.

Otherwise, determine the use and loading of each unit as follows:

Excess number of units available.

- Any such excess is the difference between the number of units of plant available and the number imposed. A certain number of units, equal to the excess, then must be excluded or shut.
- Determine which, if any, of the available units can be shut while still leaving the remainder able to meet the rate of needed transfer of energy QN plus any required reserve.

Test limits.

- If not enough units can be shut, the energy system is not feasible; stop the calculations, and return "No solution" to the calculation subprogram.

Units of plant to work.

- Begin with all the available units working; then progressively shut the excess in all combinations, excluding any that cannot be shut if it would result in overload. If their performance curves are regular and similar, it may be feasible to eliminate excess units simply by extrapolating loads.

Load each combination of working units.

Determine and assign the best set of loads for all units so as to minimize the rate of supplied transfer of energy QS for the section of plant.

- Initially distribute the total load among the units in proportion to their ratings.
- Then, if individual control is available and performance curves are "regular," apply a method of incremental loading (increase the load on the most efficient unit and decrease the load on the least, repeatedly) until the total load is distributed most efficiently, with the value of QS minimized (see "Plant Performance Curves" in Appendix C).

Best set of units to work.

- Progressively test the value of QS, and return the best result to the calculation subprogram as the set of units and loadings to adopt.

APPENDIX C

Techniques

This appendix describes some of the current techniques used for demonstrating the unified method. They are, however, just a beginning. New and improved techniques should evolve in the future as more people use and contribute to the method. For other topics, see the "Catalog of Examples" and "Notes for Programmers" in the software collection that can be downloaded from the website associated with this book, www.mhprofessional.com/TostevinEnergySystems.

Calculation

Calculation of an energy system is at the heart of processing for the unified method; it is always required, and it has to occur repeatedly, for each unit time, during each step of synthesis.

Appendix A outlines how the computer program conducts the calculation, working from the demands upstream through a sequence of cells to the sources of energy. Appendix B specifies the detailed steps of the calculation subprogram and states the prior need to define a logical *order of calculation*.

The order of calculation implied by Figure C.1 is straightforward because the energy system has just one demand, a single source of energy, and no alternatives. However, with two or more simultaneous demands or sources of energy or both, it is necessary to conduct the steps of calculation for different sequences of cells in a certain order. A sketch of the cellular view of an energy system helps to clarify this. The order of calculation depends not only on the relation of the cells to

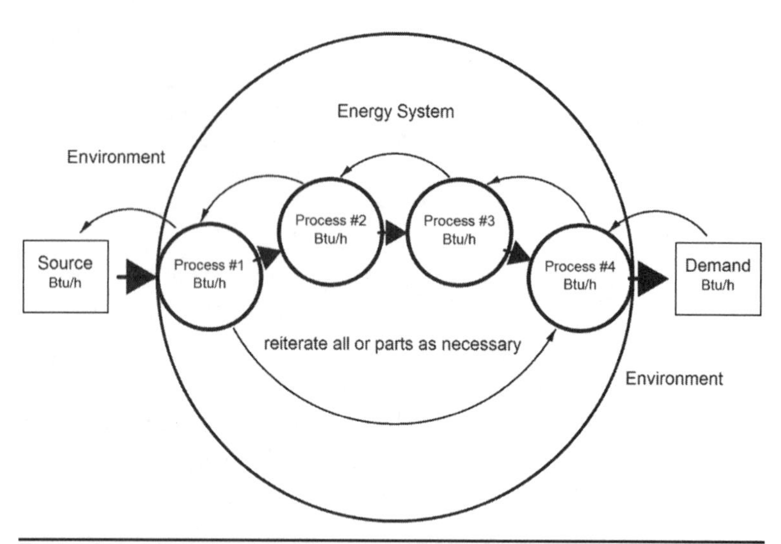

Figure C.1 Calculation of an energy system.

the external demands, and to each other, but also on any internal demands within the energy system itself. Further, the sequence in which the synthesis subprogram imposes different sections of plant should match the order of calculation.

The computer program for demonstration "Operate" (Figure 6.2) uses the following order of calculation:

```
Three external demands ND
For ND = 2 ELECTRICITY demand
  NN = 18 (Steam turbo generators)
  NN = 44 (Gas turbo generators)
For ND = 3 LP PROCESS STEAM demand
For ND = 1 HP PROCESS STEAM demand
  Boiler feedwater
  NN = 16 (Feedwater pumps)
  Condensate
  NN = 05 (Deaerators)
  NN = 03 (HP/LP pressure-reducing valves)
  NN = 01 (LP steam dump valves)
  Reiterate for auxiliary steam from NN = 16
  NN = 34 (Boilers)
  Reiterate for electricity from ND = 2
  Steam balance
  Reiterate for steam balance from ND = 2
One source of energy NE
For NE = 1 GAS FUEL
```

Similarly, the computer program for demonstration "Design" in Chapter 6 uses the following order of calculation:

```
Three external demands ND
For ND = 6 CHILLED WATER demand
  NN = 15 (Electric motor refrigeration)
  Reiterate for auxiliaries from ND = 6
  NN = 07 (Ice storage)
For ND = 5 HOT WATER demand
  NN = 02 (D/G exhaust water heaters)
  NN = 19 (Fuel-fired water heaters)
  Reiterate for auxiliaries from ND = 5
For ND = 2 ELECTRICITY demand
  NN = 17 (Diesel generators)
  NN = 08 (Electricity tie: transformers)
For NR = 1 COOLING WATER residual
  NN = 11 (Cooling towers)
  Reiterate for electricity from ND = 2
Three sources of energy NE
For NE = 2 ELECTRICITY
For NE = 1 GAS FUEL
For NE = 3 DIESEL OIL
```

Note that both these computer programs provide for a maximum configuration of plant, and hence an order of calculation, of greater scope than is required for the demonstrations in this book. See "Notes for Programmers" in the software collection.

Containers

Chapter 5 outlined how the unified method uses a set of six containers, SIZE, TT, PP, MM, EFFY, and VV, to carry information in the cells of energy systems. The information in each cell has to be complete and correct for each unit time. It may be acquired from various parts of the computer program: Some values may be fixed and given, others imposed, others calculated during the processing, others obtained from numerical tables or mathematical expressions, and some measured by instruments of an energy system that is working. The following notes explain the forms of information held in the containers.

Container SIZE

This contains the sizes of the processes and the plant represented by the cell, expressed as rates of transfer of energy. Information in SIZE is required for all processes. It includes both current values and limits of value. The following symbolic names contain information for SIZE:

SZHZ	Maximum continuous rating of a unit of plant at standard conditions
SZHI	Maximum permissible rating of the unit at current conditions
SZOP	Current load on the unit, equivalent to its rate of needed transfer of energy
SZLO	Minimum rating of the unit, the lowest load at which it can work continuously

For energy conversion processes, container SIZE usually requires a mathematical expression or numerical table to provide values when the plant is working at nonstandard conditions.

For heat exchange processes, SIZE requires information about heat transfer coefficients so that ratings can be determined for a certain surface area and for values in other containers.

The value of SZHI is the same as that of SZHZ when the plant is new and working at standard conditions. However, SZHI is less than SZHZ when the plant is worn, dirty, or damaged. SZHI can be higher than SZHZ if an overload is acceptable for a short period.

The value of SZLO is usually less than that of SZHZ but greater than zero. For example, SZLO might be 5 percent of SZHZ for a pump or 60 percent of SZHZ for a turbo compressor. SZLO is equal to SZHI for a plant that can work only at full load, on or off.

Container TT

This contains the temperatures of the processes and the plant represented by the cell, expressed as °C or °F. Information in TT is required for all thermal processes. It includes both current values and limits of value. The following symbolic names contain information for TT:

TI	Initial temperature of a process
TF	Final temperature of a process
TIMAX	Maximum permissible limit for TI
TIMIN	Minimum permissible limit for TI
TFMAX	Maximum permissible limit for TF
TFMIN	Minimum permissible limit for TF

The value of TI is lower than that of TF for a heating process and higher for a cooling process, but they are nominally equal for processes of condensation and evaporation. Values for TT may be fixed at the standard conditions or be imposed, or they may vary with the load on the plant and with conditions in the environment.

Limits for TT may have particular values, such as the highest permissible temperature for a combustion process in a furnace. Otherwise, the values for limits are likely to be set by the temperatures of connected processes or by the environment or both.

Limits for heat exchange processes are set by a minimum difference between TI and TF and by the approach of the temperatures of one process to those of another.

Container PP

This contains the pressures of the processes and the plant represented by the cell, expressed as bars or pounds per square inch (psi). Information in PP is required for all processes in working fluids. It includes both current values and limits of value. The following symbolic names contain information for PP:

PI	Initial pressure of the working fluid of a process
PF	Final pressure of the working fluid of a process
PIMAX	Maximum permissible limit for PI
PIMIN	Minimum permissible limit for PI
PFMAX	Maximum permissible limit for PF
PFMIN	Minimum permissible limit for PF

The value of PI is lower than that of PF for a compression process and higher for an expansion process, but they are nominally equal for processes occurring at constant pressure, such as condensation and evaporation. Values for PP may be fixed at the standard conditions or be imposed, or they may vary with the load on the plant and with conditions in the environment.

Limits for PP may have particular values, such as the highest permissible pressure for a condenser vessel. Otherwise, the values for limits are likely to be set by the pressures of connected processes or by the environment or both.

Container MM

This contains descriptions of the materials, most often working fluids, associated with the processes and the plant represented by the cell. Each description includes the name, nature, composition, and properties of the material, particularly its specific energy (MJ/kg or Btu/lb) and density (kg/m^3 or lb/ft^3). Such information is required for all processes involving materials. The following symbolic names contain information for MM:

WM	Literal name of the material
HI	Specific energy of the initial state of the material
HF	Specific energy of the final state of the material
DI	Density of the initial state of the material
DF	Density of the final state of the material

Values for container MM may be fixed for the standard conditions. However, MM usually includes a mathematical expression or numerical table to provide such information over a range of values in TT and PP, especially for combustion processes and those involving steam or refrigerants.

Limits for MM are set by both the characteristics of the materials and values in containers TT and PP.

In addition to calculations for the rates of transfer of energy, the processing also uses the information in MM to calculate the rates of flow of working fluids.

Container EFFY

This contains information that provides a value for the efficiency or coefficient of performance for conversion of energy in the cell over a range of load and conditions. The following symbolic name, the same as that of the container itself, contains the current value derived from that information:

EFFY Efficiency or coefficient of performance of the processes and the plant in the cell at current load and working conditions, expressed as a decimal value.

The value of EFFY expresses the ratio of the rates of needed to supplied transfers of energy at the cell. It is less than or equal to 1.0 when expressing efficiency, and typically greater than 1.0 when expressing coefficient of performance for refrigeration plant. A value for EFFY is required for all processes, even if it is only 1.0 when there is no conversion of energy.

Information from manufacturers can provide values for EFFY over a range of loads on the plant at standard conditions. However, a mathematical expression or numerical table is also required to adjust or provide such values when the plant is working at nonstandard conditions, particularly over the range of values in containers TT, PP, and MM.

This appendix explains the use of a plant performance curve, represented by a set of ABCD coefficients, to provide approximate values of EFFY at standard conditions for many kinds of plant commonly found in energy systems.

Container VV

This contains numerical values for the costs of owning and using the plant represented by the cell, expressed as follows:

VC	Capital cost of purchase, installation, and commissioning, dollars per unit of rating
VF	Annual fixed cost, dollars per year per unit of rating
VR	Running cost, dollars per hour of use per unit of rating

Each component of VV applies to that plant that is fully installed and ready to work, including the cost of foundations, minor auxiliaries, piping, ducts, controls, instrumentation, cables, and labor. Values for VF are costs such as insurance and indirect labor that apply whether the plant is used or not. Values for VR are costs such as direct labor, cleaning, and maintenance that apply to periods in which the plant is in fact working.

The essential requirement for information in container VV is that the capital and recurrent costs of the plant can be calculated:

- First for each unit of plant on its own
- Then for each section of plant by adding the costs of its units
- And then for the whole configuration of the plant by adding the costs of all sections of plant

Other components of VV, using symbolic names beginning with V, contain whatever additional information is required for evaluation, such as the space and weight of plant.

Standard Conditions and Constants

At the outset, the work of gathering information for an energy system requires a definition of certain standard conditions that will apply to sources of energy, demands, the environment, plant, and working fluids. An energy system may work only occasionally and precisely at those conditions, but they provide a permanent reference for the processing to adjust its calculations to current conditions.

The unified method requires that the same standard conditions apply throughout the whole energy system, including all options and alternatives. The standard conditions are likely to be those used traditionally at the geographic location of an energy system or by the manufacturers of its principal plant. However, it is often necessary to adjust the data from some manufacturers to those conditions, and this is done when gathering those data and converting them into universal terms.

Values for information in containers TT, PP, and MM are measured from a consistent physical datum, such as 0 °C and one bar. Such datum values are part of the standard conditions.

Once an energy system is built and working, some conditions, such as the heating values of fuels, voltages of electricity, and pressures of steam, become nominally fixed. These may become part of the standard conditions, even if they vary in practice within modest limits.

The machinery database file lists and defines the standard conditions, and all the information in that file has to be coordinated with those conditions. A utility program for the machinery file allows interactive entry of values for standard conditions, and it can reset other information in the file on changes to those conditions. The demonstrations in this book show examples.

Constants

Calculations for energy systems usually have to include several minor constants that affect the accounting for transfers of energy. Energy balances fail during the processing if such constants are omitted.

Values are required for nominal constants such as rates of boiler blow-down and other losses of working fluids. It is also appropriate to use constant values for minor internal demands, such as instrument air and lighting, that run continuously when an energy system is working regardless of the external demands.

The design of an energy system requires estimated values for all such constants based on experience with similar systems or on calculations for local conditions. Operation of an existing energy system requires such values to be determined by field tests and adjusted periodically.

The machinery file lists the values for constants, and its utility program allows their values to be entered and adjusted interactively.

Plant Performance Curves

Computer programs for the demonstrations in this book use a single curve to describe the performance of each unit of plant. The curve is a graph of the efficiency or coefficient of performance, and hence the value of EFFY, over a range of loads at standard conditions (Figure C.2).

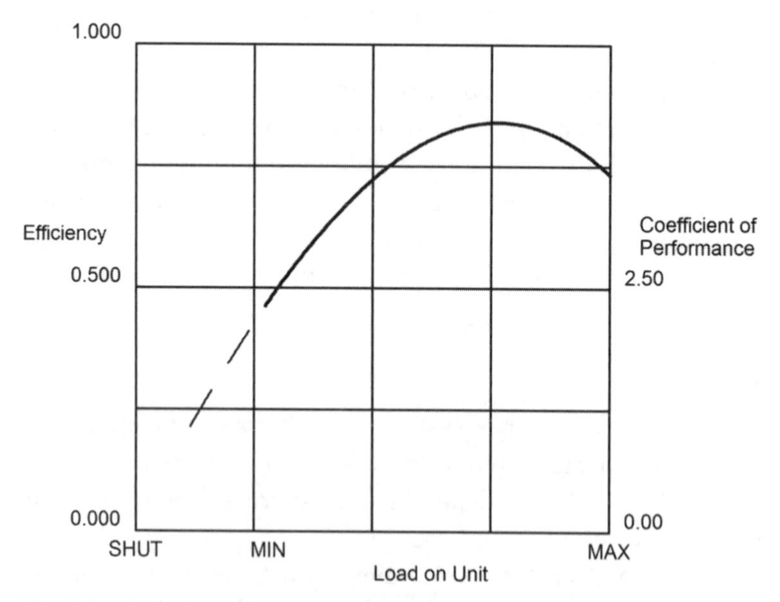

FIGURE C.2 Plant performance curve.

The shape of the curve in the figure is "regular," in the sense that it is typical for many different kinds of plant in energy systems. However, the performance of some kinds of plant is "irregular," for which the use of such a curve is not appropriate, as explained below.

It is often adequate to express a regular performance curve as the values of four (A, B, C, and D) coefficients of an optimal polynomial approximation. Moreover, it is usually possible to convert manufacturers' data for the performance of their plant at certain conditions to a set of such coefficients as follows:

- Define the standard conditions for the energy system. These might be the same as the conditions specified by the manufacturer. If not, the manufacturer still may have given enough information to enable adjustment of the data to different conditions. Otherwise, ask the manufacturer to provide data, especially for the required standard conditions.
- Define the maximum and minimum ratings of a unit of plant, at standard conditions, as values of SZHZ and SZLO, respectively.
- Express the range SZLO to SZHZ in terms of a *load ratio* R, where R = SZOP/SZHZ and SZLO/SZHZ $\leqslant$ R $\leqslant$ 1.0, where SZOP is the current load on the unit, equivalent to the rate of its needed transfer of energy QNU.
- Convert manufacturer's data, at standard conditions, into a set of values for R and EFFY distributed over the range of load from SZLO to SZHZ, where EFFY = QNU/QSU and QSU is the rate of its supplied transfer of energy.
- Fit an optimal polynomial curve to such a set of values so that the relation between EFFY and R is expressed as $EFFY = A + (B \times R) + (C \times R^2) + (D \times R^3)$.
- The values of A, B, C, and D (the ABCD coefficients) describe the performance curve for the unit of plant at standard conditions (Figure C.3).

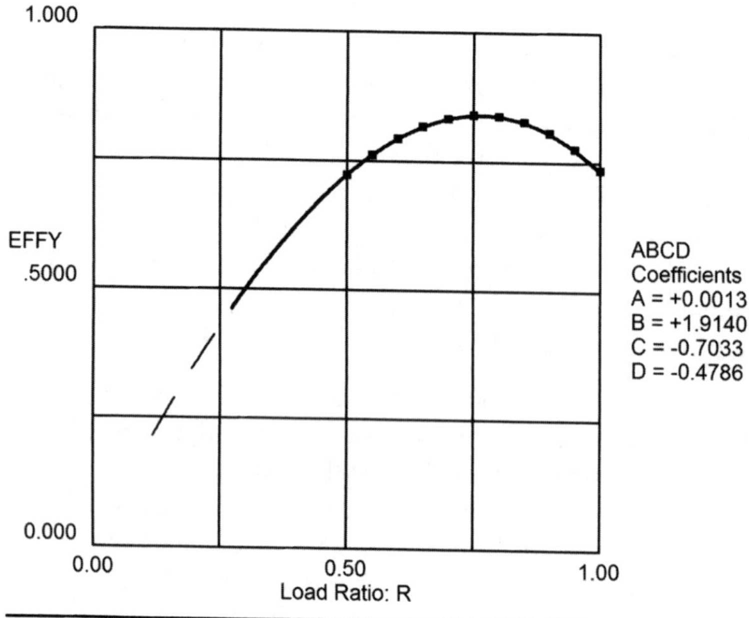

FIGURE C.3 ABCD coefficients.

When calculating an energy system, the processing uses ABCD coefficients in the following way:

- Values for SZHZ, SZLO, and the ABCD coefficients are given from the machinery file.
- The calculation subprogram determines the value of the needed transfer of energy QNU.
- SZOP is equivalent to QNU, and the value of R is SZOP/SZHZ.
- ABCD coefficients provide the value of EFFY for that value of R.
- The value of QSU is then QNU/EFFY, before any adjustments for nonstandard conditions and parasitic transfers of energy.

The machinery database files for the demonstrations in this book list the values for ABCD coefficients as part of the given information for each unit of plant. Utility programs enable people to test and change those values.

As an example, enter the data for loads and efficiencies into a curve-fitting computer program as follows:

```
R (load ratio)        EFFY (efficiency)
0.5000                0.7232
0.5500                0.7623
0.6000                0.7940
0.6500                0.8178
0.7000                0.8335
0.7500                0.8407
0.8000                0.8389
0.8500                0.8280
0.9000                0.8074
0.9500                0.7768
1.0000                0.7360
```

Then the curve-fitting program derives the following values for the ABCD coefficients and fits the optimal polynomial curve shown in Figure C.3:

```
A = + 0.0013
B = + 1.9140
C = - 0.7033
D = - 0.4786
```

Index the symbolic names of the ABCD coefficients for each unit of plant in the same way as other information, such as A[NN][NU] and B[NN][NU].

Values of the ABCD coefficients based on manufacturer's data usually describe the *design curve* for a unit of plant. Another set of values may be obtained from field tests on a unit or from the records of instrument and control systems. Such values describe the *current curve*, which is usually inferior to a design curve because of aging and fouling of the plant. Operation of an existing energy system requires current curves, although it is usual also to keep the design curves in the machinery file as a permanent reference (Figure C.4).

The performance of some kinds of plant over a range of loads can be discontinuous. The efficiencies of some kinds of steam turbines change abruptly as inlet nozzles open or shut above and below certain loads, and coefficients of performance

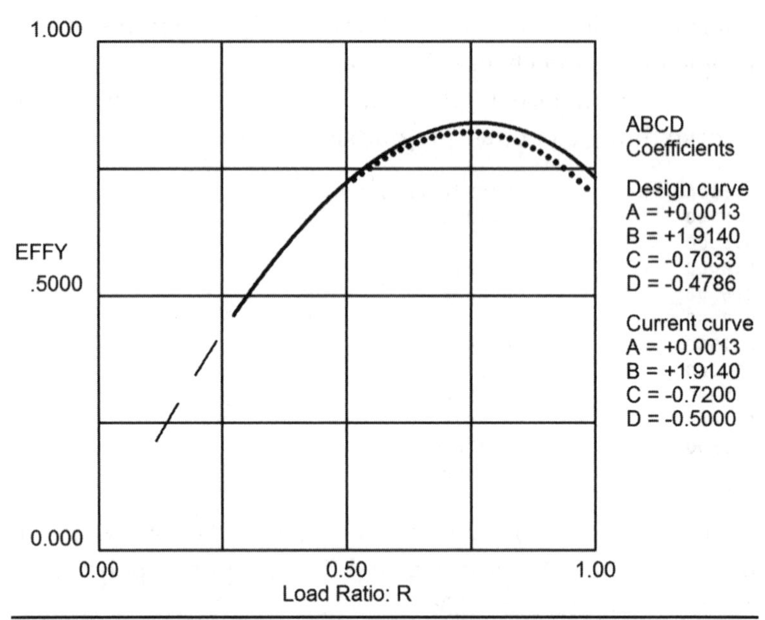

FIGURE C.4 Design and current performance curves.

of multicompressor water chillers change abruptly as one compressor starts or stops. Two or more sets of values for the ABCD coefficients then are required, one for each part of the performance curve, together with the value of the load ratio RZ at which the change occurs at standard conditions (Figure C.5).

It might be more appropriate to express the performance of some kinds of plants with a logarithmic or power curve. We still can use sets of coefficients to describe such curves, provided that they are accompanied by indicators to show the processing how to use them.

Sometimes the performance is so *irregular* that it is not possible to fit a mathematical approximation. A table of numerical data then is required, from which the processing obtains values for EFFY by a lookup technique. Further, incremental loading of such a plant is not appropriate, and a different version of the machine loading subroutine is required to evaluate all the different combinations of units in turn. See "Notes for Programmers" in the software collection.

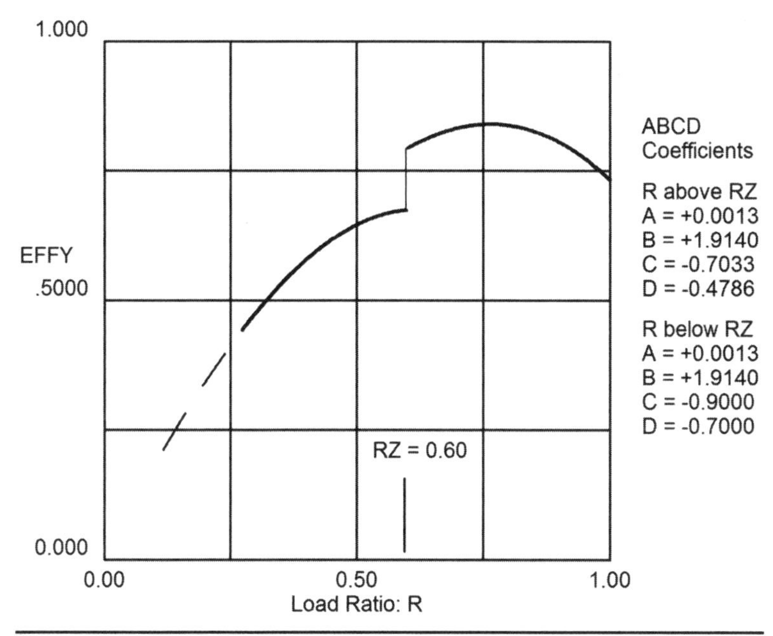

FIGURE C.5 Discontinuous performance curve.

When using a single curve for the performance of a unit of plant at standard conditions, the information in container SIZE also requires a mathematical expression or numerical table to adjust its values when the unit is working at nonstandard conditions. It may be sufficient, for example, to adjust the rating of a gas turbo generator by certain small amounts for every degree change in ambient air temperature.

Auxiliaries

In general, the working of auxiliary machines, such as pumps and fans, relates to that of the section or unit of plant they serve rather than directly to the demands on an energy system. Chapter 3 outlined three ways of accounting for the transfers of energy associated with auxiliaries (described here in more detail).

Major Auxiliaries

Large auxiliary machines may be included in the maximum configuration as a separate section of plant in their own right. The feed-water pumps of demonstration "Operate" in Chapter 6 are an example. The machinery file then includes detailed information in the containers for such plant, with which the processing directly calculates its rates of transfers of energy.

The processing accounts for the effect that parts of the transfers of energy supplied to auxiliaries may become thermal energy delivered to their working fluids.

Minor Auxiliaries

Most energy systems include several small auxiliaries that do not justify such detailed calculations; they are often motor-driven and simply run and shut down with their principal machines. A motor-driven fan for a small cooling tower is an example. When the cooling tower is working, the rate of electricity supplied to the fan motor provides the value of QAU associated with the cooling tower. The values of QAU for all such units of plant then contribute to an internal demand for electricity.

Information about the rates of transfer of energy of minor auxiliaries therefore is given rather than calculated. The demonstrations in this book express such rates as a proportion of those of their principal machines as follows:

$$\mathrm{QAU/QNU} = E + (F \times R) + (G \times R^2) + (H \times R^3)$$

where E, F, G, and H are the coefficients of an optimal polynomial approximation, and R is the current load ratio of the principal machine (Figure C.6).

A set of EFGH coefficients then is required for each unit of plant that relies on minor auxiliaries, and the processing uses those coefficients to provide values of QAU for certain values of QNU and R of its principal machine.

Two or more sets of values for EFGH coefficients are required when auxiliaries require the supply of energy in different forms, such as electricity and steam.

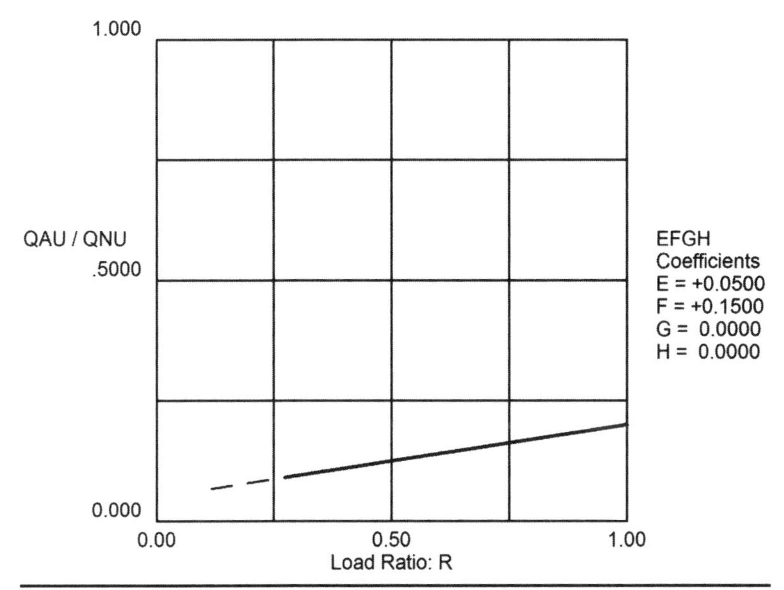

Figure C.6 EFGH coefficients for minor auxiliaries.

Packaged Auxiliaries

Manufacturers often integrate some auxiliaries with their principal machines. Engines, for example, often drive their own cooling fans and pumps. The manufacturers then usually specify the ratings and efficiencies of such packaged machines as net values, including the effects of auxiliaries. The values of QAU for such machines are zero.

Energy Storage

Chapter 3 outlined the processing of information for an energy system with storage, and Appendices A and B described that processing in more detail. This section explains the use of symbolic names for energy storage.

Index NK, a particular value of index NN, identifies storage as a separate section of plant. Symbolic names for the rates of needed, supplied, and parasitic transfers of energy for storage then are QN[NK], QS[NK], and QP[NK], respectively (Figure C.7).

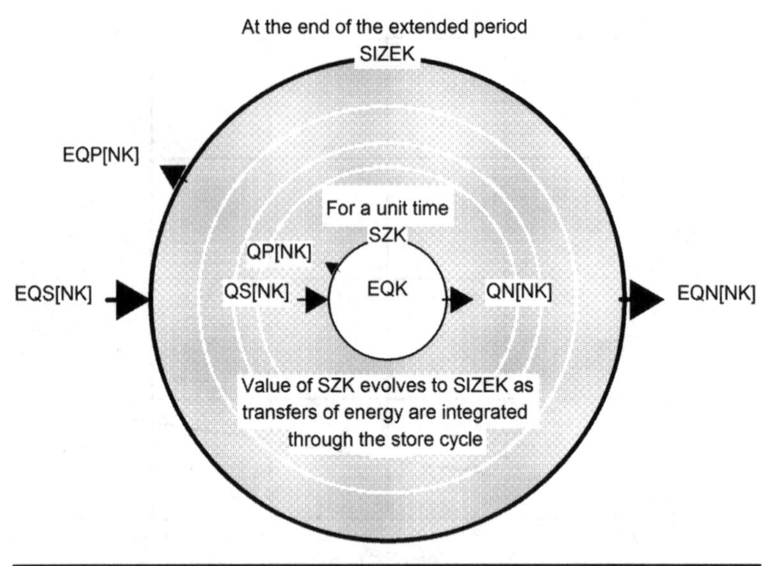

FIGURE C.7 Symbolic names for energy storage.

A residual transfer of energy usually does not apply to storage. However, storage often needs minor auxiliaries such as pumps and fans to be accounted as auxiliary transfers of energy QA[NK].

With storage included in the maximum configuration, the processing integrates the calculation for an energy system in reverse order of time through at least one store cycle. The value of energy-in-store at a particular unit time is the algebraic sum of QN[NK], QS[NK], and QP[NK] (and sometimes a portion of QA[NK]) integrated from the beginning of the store cycle.

Symbolic names EQN[NK], EQS[NK], EQP[NK], and EQA[NK] carry integrated values of the needed, supplied, parasitic, and auxiliary transfers of energy for storage, respectively. EQK carries the value of energy-in-store at any one time, whereas SZK carries the value of the equivalent physical level of storage at that time. Values of EQK and SZK are related by the conversion factor for storage CVN[NK].

For each step of synthesis, during each store cycle, the value of SZK either remains zero or rises to a certain maximum, recorded by the symbolic name SIZEK. On completion of the processing for the extended period, the value of SIZEK then represents the size of storage necessary for the energy system to remain feasible for the imposed conditions and plant. If at any time the value of SIZEK exceeds a given limit for storage, the energy system is not feasible.

Accounting for transfers of energy through the whole energy system requires that the value of energy-in-store is the same (often but not necessarily zero) at the beginning and end—datum time—of each store cycle. Again, if this does not occur and is not merely an error that shows in the energy balance, the energy system is not feasible.

The source code of the computer program for demonstration "Design" in this book includes examples of energy storage.

Generalized and Specific Information

Chapter 5 outlined the different forms of information in the machinery database file for the ratings, performance, and costs of units of plant, namely, generalized, specific, and current information. The machinery file for demonstration "Operate" in Chapter 6 contains current information. The machinery file for demonstration "Design" contains both generalized and specific information. All such forms of information provide the values required in containers during the processing.

Generalized information is merely typical of the units likely to be used in a particular energy system, and it has the same values for all units in a section of plant. Specific and current information lists the name, make, model, rating, performance, and cost for each unit individually.

The processing is virtually the same with all forms of information, and a designer may change from one to another interactively.

For generalized information, the machinery file lists values for the ABCD coefficients, the EFGH coefficients, and costs for units of plant, with suffix G appended to their symbolic names, such as AG, EG, and VCG (see Appendix G).

Generalized information usually is sensitive to the size or rating of the plant. For example, capital costs per unit of rating usually decrease as ratings increase. Adjustment of cost for rating may not be required if the values listed in the machinery file are indeed typical of the plant that is likely to be used. If necessary, however, the machinery file may list a reference size and additional coefficients to express generalized cost curves by which the processing can make such adjustments. The machinery file and utility program for demonstration "Design" show examples of these additional coefficients.

Use of Information

For design, the synthesis subprogram first allocates likely ratings and numbers of units required for each section of plant by reference both to the maximum demands on an energy system and to any interactive requests from a designer. It initially assigns generalized information to those units. Later, however, a designer has to select or at least evaluate particular makes and models of plant, for which the processing requires specific information. On request, and provided that it is available, the processing then assigns specific information in place of the generalized information.

Strictly, the processing with specific information could assemble and evaluate all the different combinations of makes and models of plant listed in the machinery file. However, part of the task of its utility program is to keep generalized and specific information coordinated, particularly for any one corporation or field of work. Solutions with both forms of information then are likely to be similar, and the results with generalized information will reduce the number of combinations needed for processing with specific information.

Results obtained with specific information may lead to commercial negotiation of prices, which often get lower in the face of competition. Initially, a designer can test the likely effect of such changes by applying adjustments interactively. Later, however, it may require a change of information in the machinery file and repeated processing until the results are finalized.

Lifecycle Cost

Demonstration "Design" in Chapter 6 uses the lifecycle cost of an energy system as a principal criterion for evaluation. A lifecycle cost combines capital and recurrent costs through the life of an energy system into a single value. The method is defined in national standard specifications (Fuller and Petersen, 1995), and the demonstration uses it in the following way:

First, the method requires values for the following parameters that represent the arrangements for financing an energy system:

LCN	Number of years (period) for which the lifecycle cost is to be evaluated, integer value, say, 1 to 40 years
LCR	Number of years (period) of life of the energy system, after which it is expected to require replacement, integer value, say, 1 to 40 years
VLCI	Discount rate, as the real annual rate of interest after the effect of inflation is removed, real value, say, 0 to 20 percent per year, expressed as a decimal fraction
VLCE	Differential annual rate of escalation of the cost of energy in relation to the rate of inflation, real value, say, –7.0 to +7.0 percent per year, expressed as a decimal fraction

The method converts the values of the preceding parameters into three multipliers (X1, X2, and X3) as follows:

```
X1 = 000.    or = LCN    or = (Y - 1.0) / (VLCI × Y)
                           where Y = (1.0 + VLCI)^LCN

X2 = 000.    or = 1.0 / Y
               where Y = (1.0 + VLCI)^LCR

X3 = LCN     or = Y * (1.0 - Z^LCN)
               where Y = (1.0 + VLCE) / (VLCI - VLCE)
               and   Z = (1.0 + VLCE) / (1.00 + VLCI)
```

The lifecycle cost of the energy system is the sum of the following:

Initial capital cost (purchase, installation, and commissioning)

+ Replacement cost (all or part of initial capital cost) × X2

+ Annual fixed cost × X1

+ Annual running cost × X1

+ Annual cost of supplies from sources of energy × X3

The lifecycle cost is expressed in present-day dollars, excluding the effect of inflation. The source code of demonstration "Design" includes a subroutine for such calculations.

APPENDIX D

Examples

The software collection that can be downloaded from the website associated with this book, www.mhprofessional.com/TostevinEnergySystems, contains example computer programs and database files for the demonstrations described in Chapter 6. The programs exist in both executable and source forms. This appendix describes the content of the collection.

No Warranty

The downloadable computer software is provided solely to explain a particular method of using information about energy systems and to illustrate that it can produce certain kinds of results. All such software is provided without warranty of any kind, either express or implied. This exclusion of warranty includes, but is not limited to, any implied warranty of completeness or merchantability or fitness for a particular purpose. The entire risk of each and any application of the method is with you, the reader.

Warnings

The computer programs are of no more than prototype quality, merely to serve as examples of the unified method, and they are neither intended nor suitable for commercial use. In particular, all the numerical and financial information in and produced by the programs is arbitrary, sufficient to demonstrate the method without necessarily applying to real practice. Even though expressed as dollars, the currency is arbitrary.

No certification or warranty is provided for the files in this collection. Download files only to an isolated destination until you are satisfied that they are safe to use on your computer.

Software Collection

Download the complete collection to your computer as described by the instructions at the end of this book.

The collection includes about 870 computer files in 74 folders, a total of 16 MB, all arranged as a local disk–based website named "EnergySystems." The website is built with Microsoft Expression Web for use with Microsoft Internet Explorer. Open, read, and click in the associated README.htm file to enter the homepage of the website. Click links on the screens to see different pages of the website (Figure D.1).

The homepage provides links to other pages. The introduction page summarizes the purpose of the unified method and the book. The demonstrations page introduces and provides access to the folders containing executable programs and their

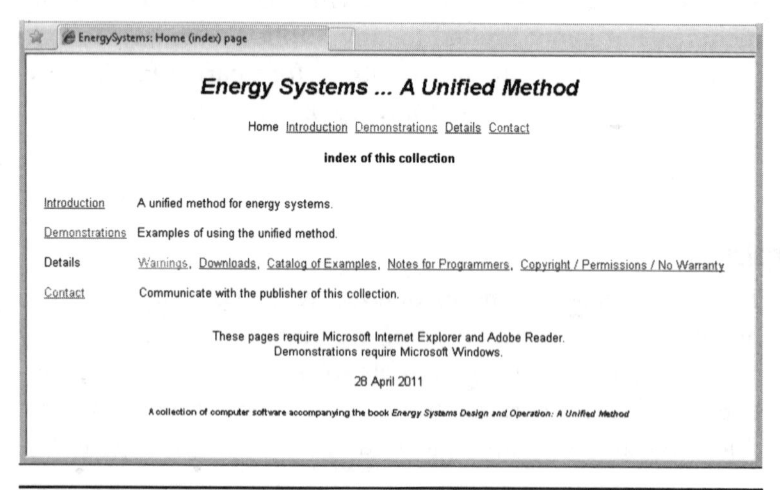

FIGURE D.1 Homepage of the local website.

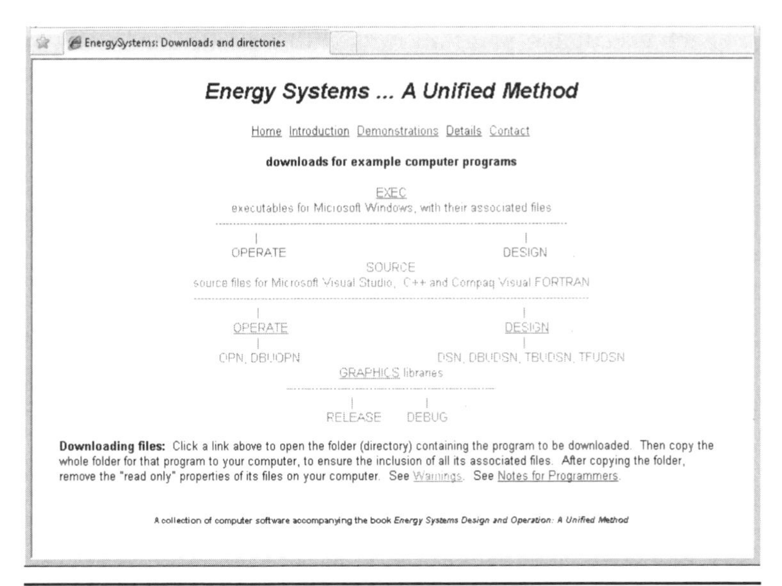

FIGURE D.2 Downloads page of the local website.

instructions. The downloads page shows and provides access to all folders containing computer programs, including those with the source code from which the executables are built (Figure D.2).

All programs are built to run with the Microsoft Windows (Win32) operating system. They do not run with other operating systems.

Executable Programs

In the EXEC\OPERATE folder, program OPN.exe is demonstration "Operate" for operating an industrial power station (see Chapter 6). This program is accompanied by a machinery database file with its utility program DBUOPN.exe.

In the EXEC\DESIGN folder, program DSN.exe is demonstration "Design" for designing the energy system for an office building (see Chapter 6). It is accompanied by machinery, time-base, and tariff database files with their utility programs DBUDSN.exe, TBUDSN.exe, and TFUDSN.exe.

Each executable program runs with a particular Job ID or jobname, such as **industry** for program OPN and **building** for program DSN. The jobname identifies the database and output files associated with the demonstration. Thus *.dbm, *.tbi, *.trf, and *.out are the names of machinery, time-base, tariff, and output files, respectively, where * is the jobname, such as industry.dbm or building.tbi.

Files in the EXEC\OPERATE and EXEC\DESIGN folders are ready to run. Each executable (.exe) program is accompanied by a (.pdf) file of instructions that can be read and printed with Adobe Reader. Certain other files are required for and accompany each demonstration, as in the following lists:

```
Folder ..\EXEC\OPERATE
Executable programs:
OPN.exe Example program OPN.
DBUOPN.exe Utility program (machinery file).
Associated files:
OPN.pdf Instructions for program OPN.
DBUOPN.pdf Instructions for utility DBUOPN.
industry.dbm Machinery database file.
industry.cfg Record of interactive dialog.
DFORMD.DLL File required for execution.
Output:
industryD.out Example output, from demands.
industryL.out Example output, from loads.
```

```
Folder ..\EXEC\DESIGN
Executable programs:
DSN.exe Example program DSN.
DBUDSN.exe Utility program (machinery file).
TBUDSN.exe Utility program (time-base file).
TFUDSN.exe Utility program (tariff file).
Associated files:
DSN.pdf Instructions for program DSN.
DBUDSN.pdf Instructions for utility DBUDSN.
TBUDSN.pdf Instructions for utility TBUDSN.
TFUDSN.pdf Instructions for utility TFUDSN.
building.dbm Machinery database file.
building.tbi Time-base file.
building.trf Tariff database file.
building.cfg Record of interactive dialog.
DFORMD.DLL File required for execution.
Output:
buildingG.out Example output, generalized info.
```

Run the executable programs either as directed from the demonstrations page of the local website "EnergySystems" or by downloading from that website to your computer in the following way:

- From the downloads page, open directory EXEC and then select the folder OPERATE or DESIGN, which contains the demonstration you wish to download and run.
- Copy that whole folder (to ensure inclusion of all its associated files) from the local website to a corresponding folder with a name of your choice on your computer.
- After copying, open the folder on your computer and remove the read-only properties of all files.
- Print a copy of the instructions (.pdf) file for the program you intend to run.
- Double-click the executable (.exe) file to start a program.
- Follow the printed instructions, and respond to messages on the interactive screens.

Source Code

Each computer program is built as an independent project in Microsoft Visual Studio with mixed programming languages, Microsoft Visual C++ 6.0 and Compaq Visual FORTRAN 6.1. Each project includes two kinds of files—those produced by Microsoft Visual Studio and those written in particular for the demonstrations in this book. Altogether, the example programs occupy more than 50,000 lines of source code.

The following SOURCE folders contain the sets of files associated with and required for each project:

OPERATE

- ..\SOURCE\OPERATE\OPN, demonstration "Operate" (~126 files, 1.22 MB)

- ..\SOURCE\OPERATE\DBUOPN, utility for the machinery database file (~104 files, 0.78 MB)

DESIGN

- ..\SOURCE\DESIGN\DSN, demonstration "Design" (~152 files, 2.10 MB)
- ..\SOURCE\DESIGN\DBUDSN, utility for the machinery database file (~108 files, 0.92 MB)
- ..\SOURCE\DESIGN\TBUDSN, utility for the time-base file (~94 files, 0.76 MB)
- ..\SOURCE\DESIGN\TFUDSN, utility for the tariff file (~86 files, 0.43 MB)

GRAPHICS Libraries

All programs except TFUDSN use local graphics libraries G2W and LIBG, each of which is also built as an independent project in Microsoft Visual Studio. The GRAPHICS folders of the software collection contain the following compiled files:

- ..\GRAPHICS\RELEASE, compiled libraries G2W and LIBG
- ..\GRAPHICS\DEBUG, compiled libraries G2W and LIBG for debug execution
- ..\GRAPHICS\graphicsU.h header file for graphics libraries

See "Notes for Programmers," which describe the content and use of the source code in more detail, below. They also tell of a ..\SUPPLEMENT folder that contains several additional examples, which may be helpful to programmers.

Specific Examples

The unified method of synthesis is a creative, building-block procedure so it instructs by examples rather than by questions and answers. The demonstration programs include a wide range of practical subjects commonly met when designing and operating energy systems, and their source code shows how the unified method deals with them. Thus they provide specific examples of code for readers who wish to write their own computer programs.

The "Catalog of Examples" in the collection identifies some 80 such examples, each with a reference by line-number to its part of the source code.

As an example, the catalog lists the subject of "energy balance" in the following way:

```
                  Program / File / Line# / Subroutine
Energy Balance
  extended period DSN / INTEGRATE.for /  554 / INTBAL()
  unit time       DSN / CALCULATE.for / 1427 / CALBAL()
                  OPN / CALCULATE.for / 1450 / CALBAL()
```

For program DSN, the file INTEGRATE.for lists the following source code at line # 554:

```
C     ENERGY BALANCE for the EXTENDED PERIOD
C     WARNING: no exclusion for deadband loading,
C       all SZLO zero this program.
C     WARNING: electricity shift excluded from balance
C       at this stage.

C     Initialize
      MBALOK = 0
C     Test energy balance
      CALL INTBAL (MBALOK)

C     Balance failed: message but NO abort.
      IF(MBALOK.EQ.0) THEN
        WRITE(WMSG,517) NV,NV
  517   FORMAT('WARNING: NV = ',I7,
     +  '\nEnergy balance failed for extended period.'C,
     +  '\nSuggest trace, examine execution NV = 'C,I7)
        CALL MSGFTN(WCAP,WMSG)

C     Verify balance for trace
      ELSE
        IF(MTRA.EQ.1) THEN
          WRITE(WMSG,521) NV
  521     FORMAT('NV = ',I7,
     +    '\nEnergy system balanced for extended period.'C)
          CALL MSGFTN(WCAP,WMSG)
        END IF
      END IF
```

The file INTEGRATE.for then lists the source code for subroutine INTBAL() arranged alphabetically among all the other subroutines it requires.

Notes for Programmers

The software collection includes the following notes at the date of publication for this book, but the version in the collection is likely to change in future as alterations and additions are made online.

The example computer programs show that the unified method works in practice, and they provide a starting point for further development and application. From that point, programmers should be free to produce programs for the unified method in their own corporate style, taking ideas, specifications, and examples from this book and projecting them into the future.

Microsoft Visual Studio

Each of the two demonstration programs and four utility programs is built as an independent project in Microsoft Visual Studio. The set of files for each project is contained in its own folder (directory) of the collection.

Most of the programs include both C++ and FORTRAN components, written for Microsoft Visual C++ 6.0 and Compaq Visual FORTRAN 6.1 respectively. Those who wish to build similar programs will need to install equivalent products on their computer, and be prepared to migrate files to those products.

When using Visual Studio, open a project's workspace to read the source files. Readers who do not have Visual Studio may read and print the source files with a text editor. Use Courier New 10 font, add line numbers, set tabs 0.5 inch for FORTRAN or 0.333 inch for C++, and save files as "plain text."

New Applications

The arrangement of programs for the unified method can be virtually the same for all energy systems, especially those within one corporation or field of work. From the outset, people may therefore design and build programs that—while still conforming to the universal structure, algorithm, and functional specification described in this book—readily allow for subsequent replication and change of detail.

For each new application, first decide on its programming language, and system of symbolic names, then create a new project in Visual Studio. Progressively select, translate, and assemble blocks of code in logical steps, make changes, and add new code. Test and verify the effect of each step before proceeding with the next. View the existing code as examples of what can and has been done, rather than what must be done.

Demonstration Programs

Chapter 6 describes two demonstration programs: OPN for *Operate* and DSN for *Design*. Each program conforms to the specifications in Appendices A and B. Each includes an Executive and four principal subprograms: Synthesize, Evaluate, Integrate, and Calculate.

The Executive part of each program combines an interactive dialog in C++ (file +Dlg.cpp, where + is the name of the project, such as DSNDlg.cpp) with a management subroutine in FORTRAN (file EXECUTIVE.for).

The principal subprograms are listed in FORTRAN files SYNTHESIZE.for, EVALUATE.for, INTEGRATE.for, and CALCULATE.for. Although similar, and of the same names, these subprograms are not interchangeable between programs OPN and DSN.

When starting executable program OPN or DSN, click Tutorial to see the maximum configuration of plant for which the program is written, and certain *special conditions* that apply to the programming. Configurations of plant selected for demonstration are less than the programmed maxima; and this is enforced by disabled controls in the screens of the interactive dialog. Later, after becoming familiar with the unified method, programmers may re-enable those controls to extend the selected configurations.

Much of the information in the demonstration programs is elementary, enough to illustrate the working of the unified method yet inadequate for use in real practice (see "Warnings" earlier in this appendix).

Programming Language

The principal subprograms, and their supporting subroutines, began and matured in FORTRAN. Although, as an exercise, they were once translated into the C language, they were then little used in that form.

The predominant use of global variables in FORTRAN suits the universal system of names for energy systems, and speeds execution. On the other hand, C++ is able to subdivide code and minimize the use of global variables in the interest of programming security. Programmers may translate the principal subprograms into whatever language they decide to adopt for the unified method. View the existing code as a beginning, and by no means the end, of programming for the method.

Components of the demonstration programs written in different languages exchange information through a MODULE (file MODULE.for), as described in the manuals of Compaq Visual FORTRAN for mixed-language programming. When building executable programs from their source code, it is usually necessary to compile that MODULE before compiling other files.

All programs in this collection that include FORTRAN components require a run-time, dynamic-link file DFORMD .DLL (a re-distributable file associated with Compaq Visual FORTRAN) placed either in folder C:\WINDOWS\SYSTEM32 of your computer or in the executable program's own folder. Copies of file DFORMD.DLL are already included in the EXEC folders of this collection.

Utility Programs

Demonstration programs OPN and DSN are both accompanied by utility programs, which are examples of those needed to create and maintain database files for the unified method.

Although similar, each utility program is tailored to suit its own demonstration program, so they are not interchangeable.

Utility programs named DBU+, TBU+, and TFU+ apply to machinery, time-base, and tariff files respectively, where + is the name of their associated demonstration program, OPN or DSN. Folders for the sets of source files for utility programs are similarly named, DBUOPN and DBUDSN.

Utility programs are mostly written in C++. However, some include FORTRAN components, and therefore (like the demonstration programs themselves) require a MODULE for the exchange of information.

The name TBU indicates a utility program for a file of time-base information for *each hour of a typical day for each month of a year* (total 288 hours). Demonstration DSN uses such a time-base file, and then annualizes its results. Such a file can adequately provide time-base information for an energy system with regular operations, and with annual climatic changes at a fixed location, such as a city building.

A similar utility program, named YBU, manages a file of time-base information for *each hour of every day for each month of a year* (total 8760 hours). Such a file provides time-base information for an energy system with irregular operations, and at different locations throughout the year, such as a ship.

The SUPPLEMENT folder of the software collection contains a set of executable and source files for a particular version of utility YBU. It is not used for the demonstrations of this book, but is included here as an example for programmers.

Database Files

Program OPN uses a machinery database file. Program DSN uses machinery, time-base, and tariff database files. They are all "comma-separated" ASCII files, which can be opened in a text editor. However, it is more appropriate to use the utility programs to edit database files.

Information in the database files of this collection is only suitable for the demonstrations, and in many places uses merely "dummy" values. Users of the unified method have to create and maintain their own stock of information in such files.

Users' Instructions

Written and illustrated instructions for the use of each demonstration and utility program are included with the sets of executable files in the EXEC folders of this collection. See +.pdf files, such as OPN.pdf and DBUOPN.pdf. As an example, this appendix includes a copy of the instructions for program DSN.

Interactive Screens

The interactive dialog for each demonstration and utility program (files +Dlg.cpp, such as DSNDlg.cpp) comprises a sequence of input and output screens named Screen A, Screen B, and so on, as illustrated in the instructions for the use of the programs. Files dlgA.cpp, dlgB.cpp, etc., contain the source code for those screens.

Output Files

Demonstration programs normally produce an output file named *.out, where * is the Job ID (or jobname) such as **industry** for program OPN and **building** for program DSN. Those files record a summary of the given information, and the detail of solutions obtained from the processing. See the Users' Instructions for each program.

Program DSN, for the design of an energy system, also normally produces the following output files: *.alt, to record the identity and values of evaluation criteria for each alternate solution, and *RESULTS.out, to record a sorted list of the alternate solutions. Note that the format of output files is *space separated*, rather than *comma separated*.

Folders named REF* contain sets of *reference* output files for the demonstration programs. Microsoft WinDiff is useful to compare such references with the content of new but similar output files.

Utility programs normally produce a backup copy of the initial database file, an updated version of that file after making changes, and, on request, a print file. However, on request when starting a utility program, it may instead create a new, relatively empty, database file. See Users' Instructions for the utility program.

All programs in this collection initially have the writing of files suppressed. In the source code of their interactive dialog (file +Dlg.cpp), the global variable MDEM is set to TRUE. Later, when familiar with the method, programmers may set MDEM to FALSE, to enable normal writing to files.

Graphics Libraries

The demonstration and utility programs require prior installation of certain graphics libraries on your computer. These libraries are provided merely for the purpose of demonstration. Programmers should build new applications to use their own corporate standard graphics.

Graphics subroutines for the demonstration programs were originally written to use the Advanced Display Graphics Support Library (GSL), issued with the early AIX operating system of IBM. However, that GSL is not required here. Instead, this collection includes a local graphics library, G2W.lib, to translate calls (similar to those for GSL) directly into functions for Microsoft Visual C++.

This collection also includes another local graphics library, LIBG.lib, to provide a few functions frequently used by the demonstration and utility programs.

Each local graphics library (G2W and LIBG) is also built as a separate project in Microsoft Visual C++. Folders named GRAPHICS\RELEASE and GRAPHICS\DEBUG contain *compiled* 'Release' and 'Debug' versions of these libraries.

The GRAPHICS folder also contains a copy of header file graphicsU.h.

Graphics libraries G2W and LIBG are *statically linked* to executable programs in the EXEC folders of this collection.

When using Microsoft Visual C++, note that the settings for such static linking (Project | Settings | Link | General) expect that copies of G2W.lib and LIBG.lib are located in folders C:\LIBCOM\Release and C:\LIBCOM\Debug on your computer, respectively. To build the executable programs it is therefore necessary to either:

- Copy the compiled versions of G2W.lib and LIBG.lib to new folders on your computer named C:\LIBCOM\Release and C:\LIBCOM\Debug, or alternatively
- Change the above settings in Visual C++ if you intend to use folders of different names for these graphics libraries

Similarly, a setting in Visual C++ (Project | Settings | C/C++ | Preprocessor) expects that a copy of header file graphicsU.h is located in folder C:\LIBCOM on your computer. To build executable programs it is therefore necessary either to copy that header file to the C:\LIBCOM folder on your computer or to change the setting.

Machine-Loading Subroutine

A key common element of programs OPN and DSN is the FORTRAN machine-loading subroutine MACHLD (see "Supporting Subroutines" in Appendix B).

The use of MACHLD depends on the performance curves of units of plant being *regular* (see "Plant Performance Curves" in Appendix C). Programs OPN and DSN, and utility programs DBU+, check that regularity with subroutine CHKABC.

For a section of plant that includes units with *irregular* performance curves, a different version of the machine-loading subroutine is required. On request, it evaluates each alternative combination of units in turn. File machlX.for, in the SUPPLEMENT folder of this collection, is an example, once required for a particular section of plant in an industrial application. It is not used for the demonstrations of this book, but is included here as an example for programmers.

Steps and Increments

Appendices A and B describe how each step of the synthesis subprogram imposes a different configuration of plant. It does so by applying increments of change to the physical size, or rating, or number of units, for each section of plant in turn.

The demonstrations of this book provide for a maximum of ten units for each section of plant, each of which may be of different rating; and they apply increments of change as the number of units to use in each section of plant. For example, the synthesis subprogram might first impose 10 units, then 9, then 8, and so on.

In addition, the machine-loading subroutine applies increments of change to increase or decrease loads on units in a section of plant. For the demonstrations of this book, these increments are of the order of one, or one half, percent of the total load on the section of plant.

For each demonstration program, its FORTRAN subroutine CONST sets the sizes of these steps and increments.

Documentation

Chapter 5, Appendix A, and Appendix B provide the principal documentation of the computer program for the unified method. Appendix C explains some of the particular techniques used for programming the present demonstrations.

Open the SOURCE folder for each demonstration or utility program to see its lists of files, and open any one of those files to see its detail. Files of source code include explanatory comments in English at each step.

Program header files include lists of subroutines, functions, and values for numerical indices. For program DSN, see files enedsn.h and ENEDSN.fi. For program OPN, see files eneopn.h and ENEOPN.fi.

Header files also contain a dictionary of symbolic names for each program. They conform generally, though not necessarily strictly, with the example dictionary in Appendix G.

Use the "Catalog of Examples" in the software collection to find (in the source code of the demonstration programs) the way the unified method deals with particular subjects.

Study the sequence and flow of programming in real time by compiling and running programs in the debug mode of Microsoft Visual C++, after placing breakpoints at entries to subroutines. First study program OPN, with its single source of energy, three demands, and single configuration of plant. Later study program DSN, with its much wider scope.

Name Changes

In the past, rates of transfers of energy were once named *energy functions*, containers were known as *functional parameters*, while MM and EFFY were named CC and PATH. Please be aware that such old names may still occur in some parts of the files of source code.

In particular, all units of plant were once named *machines*, and this term is intentionally retained in the utility programs for machinery database files.

Example Instructions

The software collection includes instructions for each of the six demonstration and utility programs. The following is an example for program DSN.

Computer program DSN synthesizes the design of an energy system for an office building, as described in Chapter 6 (see "Warnings" earlier in this appendix).

The program is built to run under the Microsoft Windows (Win32) operating system, with displays of 1280 × 800 resolution. A demonstration program may be run directly from a CD because its normal writing of external files is suppressed.

Verify that the executable program DSN.exe, the machinery database file *.dbm, the time-base file *.tbi, the tariff file

*.trf, and the program configuration file *.cfg exist in the current working directory, where * is the Job ID such as **building**.

The file DFORMD.DLL is also required in the current directory if that file is not already installed in the computer's C:\WINDOWS\SYSTEM32 directory.

The machinery file contains information about the units of plant available for design (names, ratings, performance coefficients, and costs) together with the values for their standard conditions of operation and certain constants. Maintain that information with utility program DBUDSN.

The time-base file contains information about:

- Demands on the plant for steam, electricity, hot water, and chilled water, in any combination
- Limits on the supply of fuels or electricity
- Solar radiation
- Ambient air temperatures, for each of 24 hours of a typical day for each month of the year

Maintain that information with utility program TBUDSN.

The tariff file contains information about alternative tariffs available for the supply of fuels and electricity. Maintain that information with utility program TFUDSN.

Executable program DSN.exe normally uses the program configuration file to record interactive entries to the dialog, such as options for the use of the program, sources of energy, demands, units of plant, financial values, and numerical adjustments. When restarting the program for the same Job ID, it will then resume with those entries. For demonstration, however, rewriting of program configuration files is suppressed.

Double-click DSN.exe or the green icon *DSN* to start the program. The following pages explain the sequence of interactive screens, and the entries they require.

Initial Screen

Verify that the name and date of the program are those you expect; otherwise click Cancel to stop (Figure D.3).

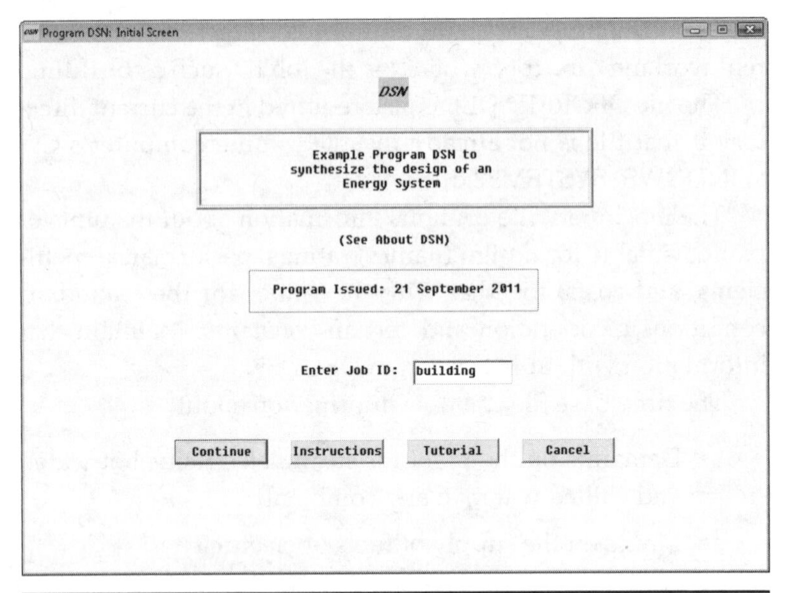

FIGURE D.3 Program DSN: Initial Screen.

Click Instructions to display an overview of how to use the program; or Tutorial to display a series of graphics screens that illustrate the purpose of the program, the configurations of plant available for design, and any *special conditions* that might apply.

Enter the Job ID, such as **building**, which identifies the machinery, time-base, tariff, and program configuration files with which you are working. The program verifies that those files exist before proceeding further.

Click Cancel to stop the program; or Continue to display the Options Screen.

Options Screen

Click radio buttons to select any combination of several options for the use of the program (Figure D.4). Some options may be disabled, indicating that the particular version of the program you are using is not equipped to deal with them. Select an initial objective for optimization (see Appendix B).

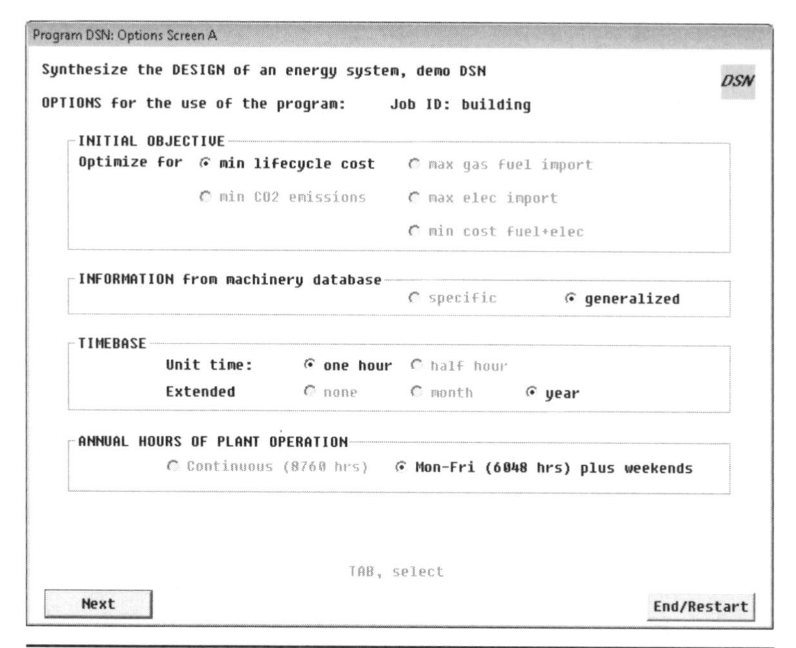

Figure D.4 Program DSN: Options Screen.

Click End/Restart to return to the Initial Screen; or Next to display the Source and Demand Screen.

Source and Demand Screen

Click check boxes to select the sources of energy that will be available to the plant and the demands that will be made on the plant (Figure D.5). Certain minimum sets of sources and demands will be required, and messages appear if necessary to help you manage the selections.

Click Back to retain the selections you have made and return to the Options Screen; or Next to adopt the selections and display the Configuration Screen.

Configuration Screen

Tab to select the maximum configuration of plant available for design (Figure D.6). Some sections of plant may be disabled

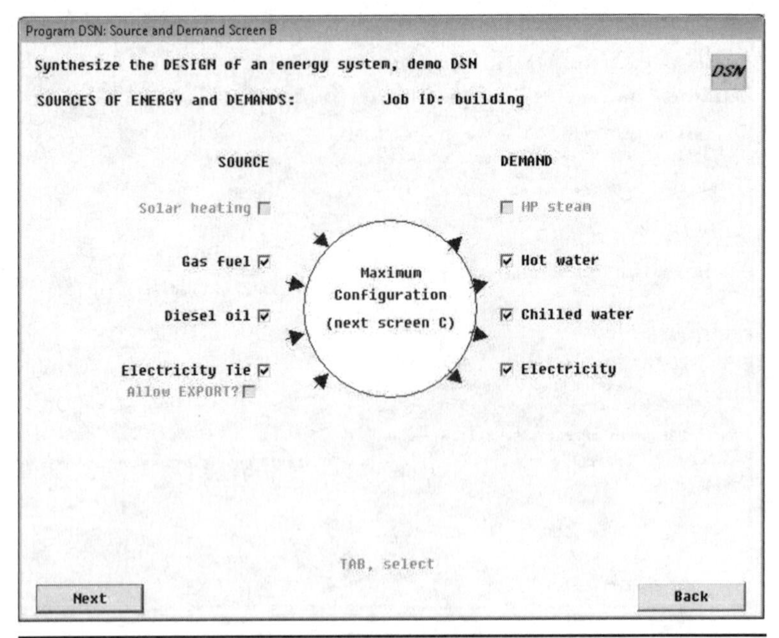

Figure D.5 Source and Demand Screen.

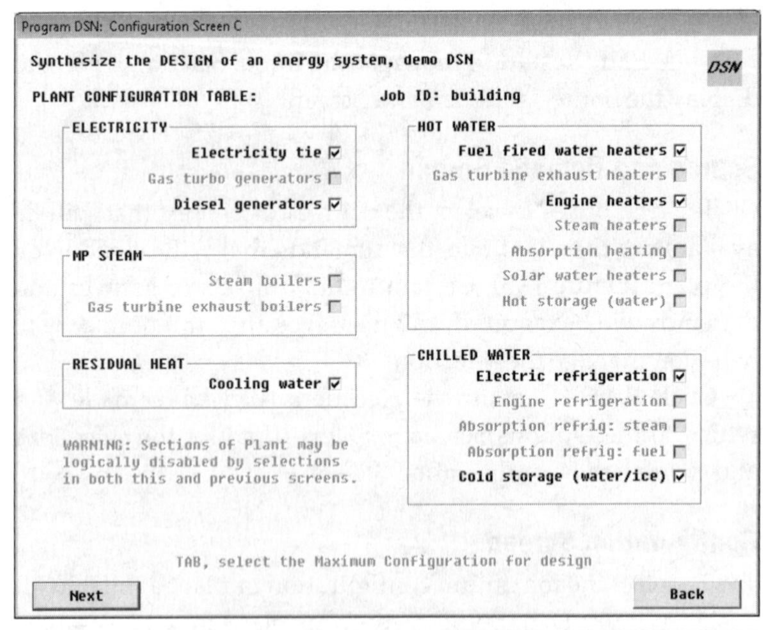

Figure D.6 Program DSN: Configuration Screen.

according to selections made on both this and the preceding Source and Demand Screen.

Click Back to retain the selections you have made and return to the Source and Demand Screen; or Next to display a graphics diagram of the selected Maximum Configuration, from which you may click Back if necessary to make changes to your selections.

Click Next on the graphics diagram to adopt the selections and display the Financial Screen.

Financial Screen

TAB to enter or change the values of Lifecycle Cost parameters that are to be used for design (Figure D.7).

Click Back to retain the changes you have made and return to the Configuration Screen; or Next to adopt the changes and display the Source and Demand Adjustment Screen.

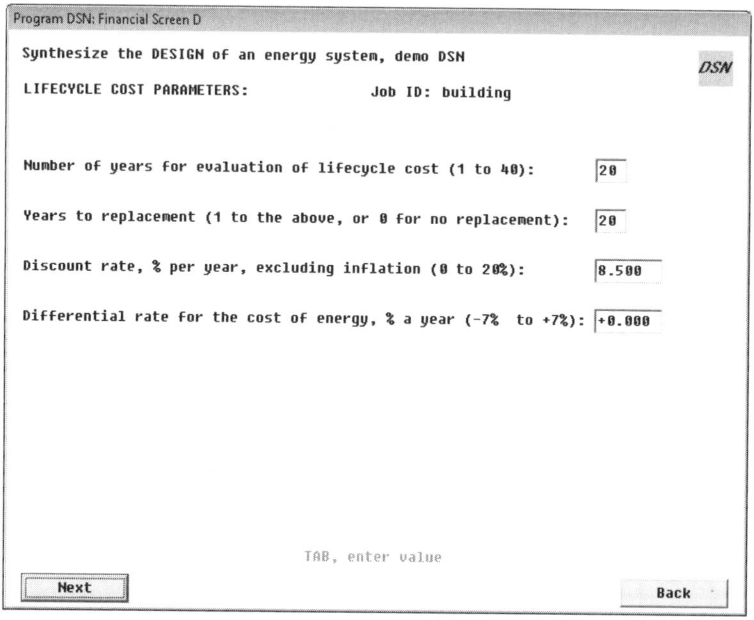

Program DSN: Financial Screen D

Synthesize the DESIGN of an energy system, demo DSN

DSN

LIFECYCLE COST PARAMETERS: Job ID: building

Number of years for evaluation of lifecycle cost (1 to 40): 20

Years to replacement (1 to the above, or 0 for no replacement): 20

Discount rate, % per year, excluding inflation (0 to 20%): 8.500

Differential rate for the cost of energy, % a year (-7% to +7%): +0.000

TAB, enter value

Next

Back

FIGURE D.7 Program DSN: Financial Screen.

Source and Demand Adjustment Screen

This screen enables you to test the sensitivity of a design to minor changes in given values for sources of energy and demands (Figure D.8).

Adjustments to the costs of sources of energy are *multipliers,* applied to costs of fuels and electricity in the tariff file (*.trf). An entry of 1.00 makes no change to the values in that file.

Adjustments to demands are *additions* or *subtractions,* applied (+ or –) to the values of demands contained in the time-base file (*.tbi). An entry of zero (+0) makes no change to the values in that file.

Fields are disabled for sources of energy and demands not previously selected on the Source and Demand Screen.

TAB to enter or change the values that will be applied throughout your use of the program. Messages appear if entries are out of range.

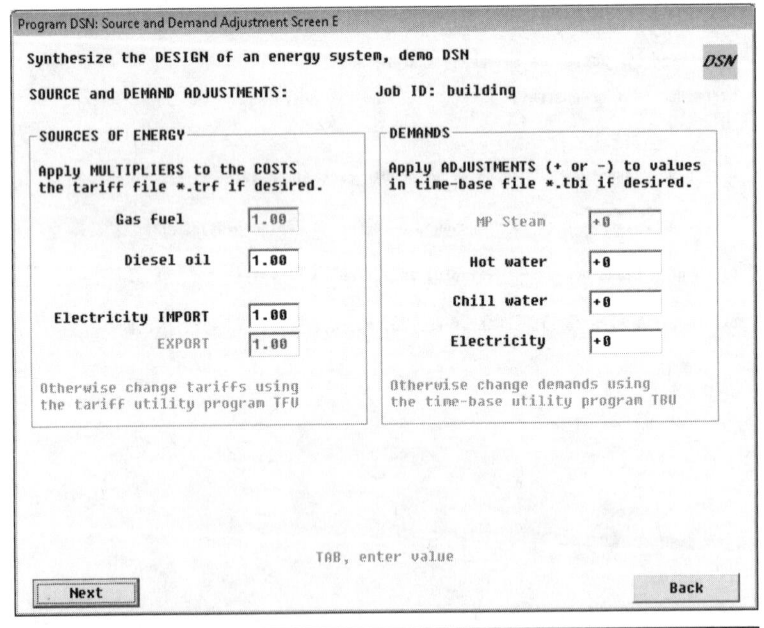

FIGURE D.8 Program DSN: Source and Demand Adjustment Screen.

Click Back to retain the changes you have made and return to the Financial Screen; or Next to adopt the changes and display the Cost of Machinery Adjustment Screen.

Cost of Machinery Adjustment Screen

This screen enables you to test the sensitivity of a design to minor changes in the given information for the costs of plant (Figure D.9).

Adjustments to the costs of units of plant are *multipliers,* applied to the costs listed in the machinery database file (*.dbm). An entry of 1.00 makes no change to the values in that file. Fields are disabled for sections of plant not previously selected on the Configuration Screen.

TAB and enter or change the values that will be applied throughout your use of the program. Messages appear if entries are out of range. Click Back to retain the changes you have made and return to the Source and Demand Adjustment

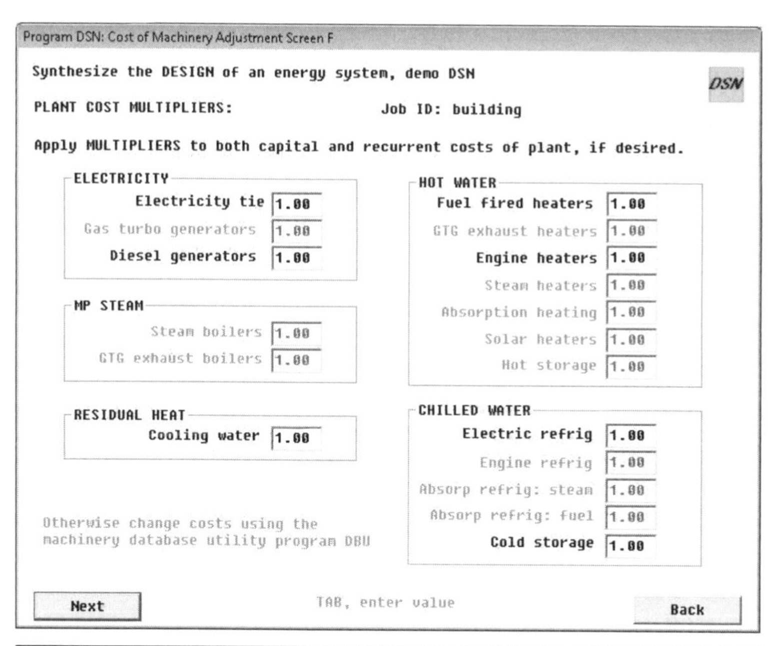

FIGURE D.9 Program DSN: Cost of Machinery Adjustment Screen.

Screen; or Next to adopt the changes and display the Current Information Screen.

Current Information Screen

Tab to enter numerical values for the maximum numbers of units of plant to use for design, whether synthesizing from either Generalized or Specific Information (Figure D.10).

Click check boxes to allow/disallow certain restrictions on the use of some sections of plant. Fields are disabled for sections of plant not previously selected on the Configuration Screen.

Enter values for any limits you wish to apply to the total capital cost of the design or its annual emissions. Zero entries apply no such limits.

Tab to enter or change the values that will be applied throughout your use of the program. Messages appear if entries are out of range.

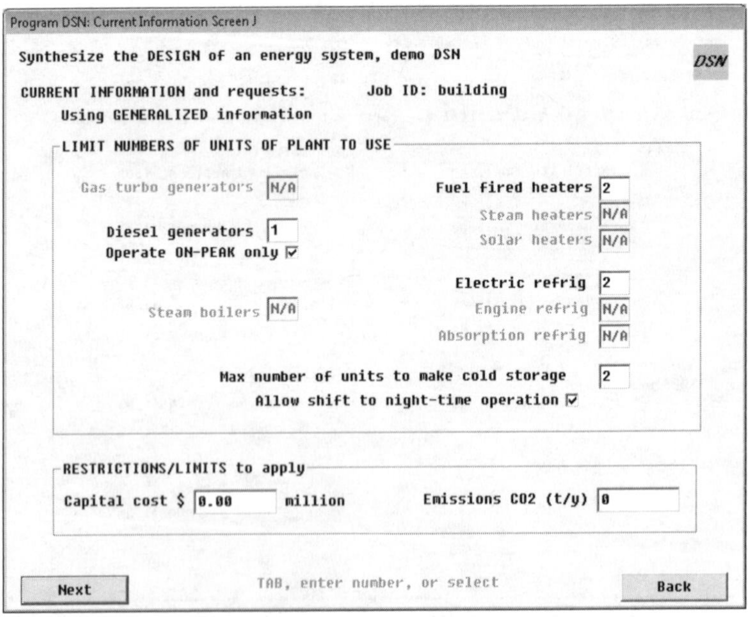

Figure D.10 Current Information Screen.

Click Back to retain the changes you have made and return to the Cost of Machinery Adjustment Screen; or Next to adopt the changes and display the Execution Screen.

Execution Screen

Click Execute to start the calculations by the program, which will use the information given in the machinery, time-base, and tariff files for the particular Job ID, and that given through the dialog screens (Figure D.11). Await and respond to the subsequent messages displayed on the Execution Screen. The progress of the synthesis is indicated in the lower part of the screen, along with the value of index NV for the current step of synthesis.

The normal final result is a set of Output Screens headed "optimal solution," the one associated with the inital objective selected from the Options Screen.

If the synthesis fails, make a note of the message obtained from the subsequent diagnosis. The failure might then be corrected by an appropriate change in some of the given information, followed by re-execution.

Provided the synthesis is satisfactorily completed, click Alternates on the Execution Screen to examine a summary of

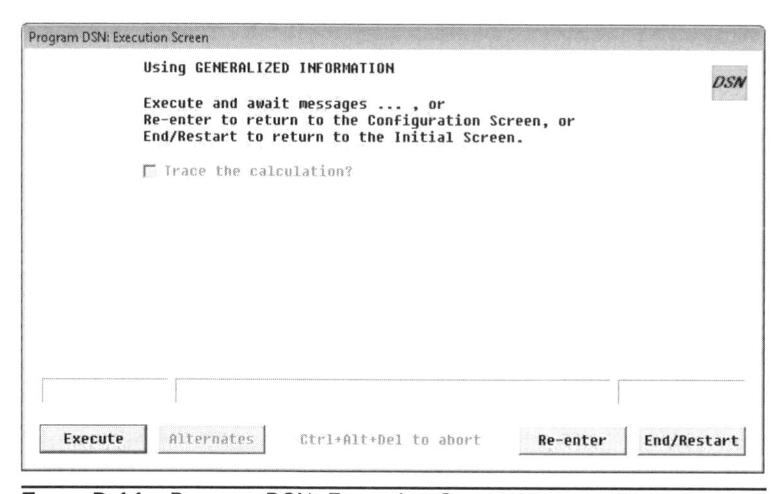

Figure D.11 Program DSN: Execution Screen.

all the different designs produced by the program. Select any one design to resynthesize and examine its detail (see "Alternates Screen" below).

On the Execution Screen, click the check box for a Trace, and respond to the subsequent messages, to follow the detailed calculations of the program. This feature is disabled for a demonstration but is otherwise intended for people familiar with the unified method. Single executions (identified by the value of their index NV) may be traced independently.

Press the CTRL + ALT + DEL keys together to stop the program if you become overwhelmed with detailed messages during a trace. Otherwise stopping from the trace may produce a set of Output Screens headed "solution in trace." Take care not to use such output as optimized information.

Click End/Restart to return to the Initial Screen.

Output Screens

Several screens are displayed on completion of the execution of the program (Figure D.12). The screen shown is REQ, a summary of the required ratings of plant in tabular form. Click REQ again (or Next) to display similar information alternately in tabular and graphics form.

Click ENE to display an annual energy summary for the design, alternately in tabular and graphics form; or $ to display information about the costs of sources of energy and plant; or DMD to display a summary of the demands on the plant; or ENV for a summary of the conditions in the environment.

Click other buttons to display information about particular sections of plant, some alternately in tabular and graphics form, as follows:

- ELE (electricity), gas turbo generators, diesel generators, electricity tie
- STM (steam), boilers, and gas turbine exhaust boilers
- HW (hot water), boilers, steam heaters, engine exhaust heaters, hot water storage

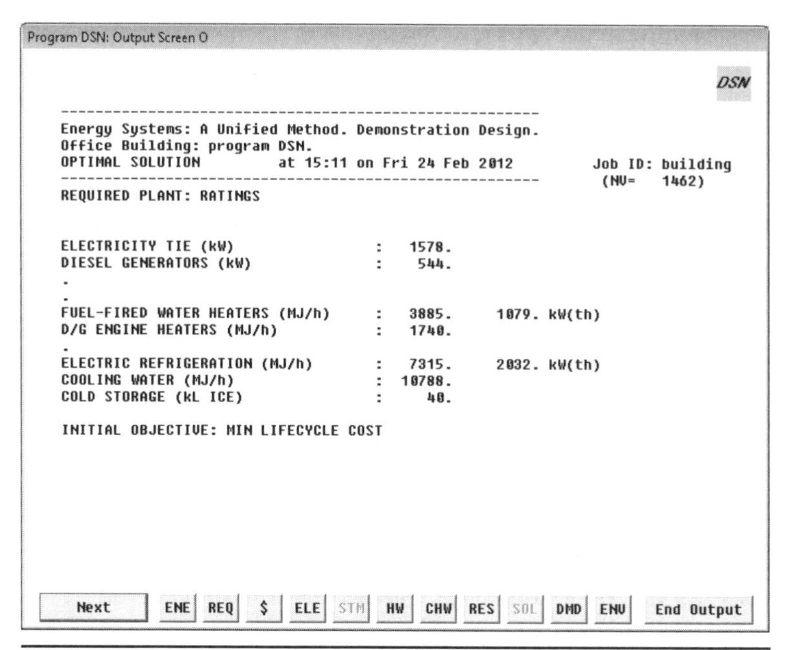

FIGURE D.12 Program DSN: One of the Output Screens.

- CHW (chilled water), electric, engine and absorption refrigeration, cold storage
- RES (residual heat), cooling water
- SOL (solar), water heaters

All the above and some additional output information is normally written to an output file, although this is suppressed for a demonstration.

Click End Output to return to the Execution Screen.

Alternates Screen

The Alternates Screen is accessed by clicking Alternates on the Execution Screen, effective only after satisfactory completion of a synthesis (see Figure 6.7).

Answer a question, yes or no, if necessary to reduce the number of different designs to be displayed.

Usually, many different designs will have been synthesized for the given information. The Alternates Screen displays a summary of those designs after they have been sorted into numerical order according to their Lifecycle Cost. The summary identifies the designs that correspond to several different criteria, such as those of maximum and minimum capital cost.

Click Plot to see a graphic display of the Lifecycle Cost of all the different designs. Enter any valid "design number," then click Execute to resynthesize that design and see its details on the Output Screens.

Press End Alts to end your examination of Alternate Solutions and return to the Execution Screen.

End Example Program DSN

To end your use of the program, click End/Restart on the Execution Screen, then click Cancel on the Initial Screen.

Normally the program will have written an output file (named *.out where * is the Job ID), containing both a summary of the given information and the set of tabular output screens produced by the most recent execution. For a demonstration, however, writing the output file is suppressed.

The program also normally writes additional output files that list summaries of the solutions obtained by the processing. An alternates file (*.alt) lists all solutions. A results file (*RESULTS.out), derived from the alternates file, lists solutions ranked by their value for a particular evaluation criterion. For demonstration, however, alternate solutions are recorded in an internal array rather than a file, so that the program can be run from a CD. If output, alternates, and RESULTS files are available, examine and print them with a text editor such as Microsoft WordPad.

Unless a demonstration, re-execution of the program from the Initial Screen will remove existing output files and write new files.

Examine or print the information in the database files by using utility programs DBU(DSN), TBU(DSN), and TFU(DSN).

Errors and messages that stop the program may indicate that the required machinery, time-base, tariff, or program configuration files are missing or corrupted. If so, add or replace those files with copies known to be sound, and run the program again.

APPENDIX E

Definitions

Terms used to describe the unified method for energy systems in this book and its example computer programs.

ABCD coefficients A set of four (A, B, C, and D) coefficients, the values of which define an optimal polynomial approximation of the performance curve of a unit of plant at standard conditions.

Auxiliary Plant or equipment that supports the operation of a principal (usually larger) unit or section of plant—major, minor, packaged auxiliaries.

Boundary A notional borderline that separates the region occupied by the processes and the plant of an energy system from the region of their surroundings.

Calculation Determine the values of all components of information necessary to describe the operation of a particular energy system in a certain unit time during which conditions are notionally constant.

Cell Encapsulated view of the processes or group of associated processes, plant, and transfers of energy in a particular part of an energy system.

Coefficient of performance *See* Efficiency.

Concept A particular kind of energy system, distinguished by its combination of sources of energy, plant, and demands.

Configuration A particular physical arrangement of the plant of an energy system—maximum, imposed, current configuration.

Container One of six "boxes" of information carried by the cells of energy systems, the set of which is the same for all cells—SIZE, TT, PP, MM, EFFY, and VV.

Database A computer file of information arranged in the universal terms required for the unified method—machinery, time-base, and tariff files.

Datum time A particular unit time, which represents the beginning and end of a store cycle for the use of energy storage.

Datum values Defined starting points for the measurement of values for parameters such as temperature and pressure.

Demand A rate of transfer of energy of particular form required from and delivered by an energy system at a particular time—maximum, average, current demands.

Design Production of engineering plans and specifications necessary for building and commissioning an energy system.

Efficiency Value of the ratio of the rate of needed to the rate of supplied transfer of energy for the processes and plant represented by the cell of an energy system. For a refrigeration plant, efficiency is expressed as coefficient of performance.

EFGH coefficients A set of four (E, F, G, and H) coefficients, the values of which define an optimal polynomial approximation of the ratio of the rate of needed transfer of energy for a minor auxiliary to the rate of needed transfer of energy of the principal plant with which it is associated.

Energy-in-store A colloquial term for the effective content of energy storage at a particular time.

Energy storage A particular section of plant that accumulates transfers of energy "in" at some times and releases similar transfers of energy "out" at other times, all within the period of a store cycle.

Energy system A combination of processes and plant designed, built, and operated to provide electricity, steam, propulsion, heating, cooling, and other services for public utilities, process industries, ships, factories, and buildings.

Environment The region outside the boundary of an energy system, usually expressed in terms of its prevailing physical, biologic, and social conditions.

Evaluation Determine a value for each of several attributes and consequences of the design and operation of a particular energy system.

Extended period A defined, relatively long period of time that extends for a succession of unit times—day, week, month, year.

Integration Progressively add, through a succession of unit times, the values of certain components of information describing the operation of a particular energy system in each unit time.

Load on plant The rate of needed transfer of energy required from a unit or section of plant at current conditions.

Load ratio A value that defines the load on a unit of plant as a proportion of its rating.

Objective A defined criterion or combination of criteria by which a process of optimization can determine the best of several different energy systems.

Operation The day-to-day running and management of an energy system after it is built and commissioned.

Optimization A procedure that finds the best of a number of different alternatives for the concept, design, and mode of operation of an energy system.

Order of calculation A logical sequence of demands, sections of plant, and sources of energy in which to conduct the calculations of an energy system.

Performance curve A graph of the value of efficiency or coefficient of performance for a unit of plant over a range of loads at standard conditions.

Plant Manufactured machinery and equipment used to contain, connect, and control the processes of energy systems.

Process of an energy system A natural occurrence characterized by a change of state of material and certain transfers of energy.

Rating The maximum load that a unit of plant can maintain continuously at standard conditions.

Section of plant A group of several units of plant, each of which can contain and control similar processes and provide similar transfers of energy.

Source of energy A rate of transfer of energy of particular form required by and supplied to an energy system at a particular time.

Standard conditions A defined set of values for conditions of the environment and conditions of the processes of an energy system, which provide a fixed reference for determining rates of transfers of energy and ratings of plant.

Store cycle A period of a succession of unit times during which the value of energy-in-store rises and falls back to the same value.

Symbolic name A short alphanumeric combination that identifies a particular component of information for energy systems, necessary for computer programming.

Synthesis Systematically assemble information about the conditions and elements of energy systems into different combinations, each of which defines the conditions a particular energy system and presents it for calculation, integration and evaluation.

Unified method A way of conducting engineering for the concept, design, and operation of all kinds of energy systems.

Unit of plant A single machine or "package" of equipment, the processes of which provide a required set of transfers of energy.

Unit time A defined, relatively short period of time during which conditions in the environment and conditions of the processes and plant of an energy system remain notionally constant—second, minute, hour.

Universal terms An organization of information that is the same for all energy systems and which is necessary for a unified method.

Utilization of plant The extent to which a unit of plant is actually used during an extended period, usually a year, expressed as a proportion of its maximum capacity for such use.

APPENDIX F
Abbreviations

Used in This Book and Its Example Computer Programs

AVL	available (unit of plant)
bar	one standard atmosphere pressure
Btu	British thermal unit
°C	degrees centigrade
D/G	diesel generator
DB	dry bulb (air temperature)
°F	degrees Fahrenheit
ft^3	cubic foot
GJ	gigajoule
GJ/h	gigajoules per hour
GTG	gas turbo generator
GTGX	GTG exhaust heater/boiler
HP	high pressure
kg	kilogram
kL	kiloliter
kW	kilowatt (electrical)
kWh	kilowatt-hour (electrical)
lb	pound mass
LP	low pressure
m	meter
m^2	square meter
m^3	cubic meter

MB	megabyte
MJ	megajoule
MJ/h	megajoules per hour
MVA	megavolt-ampere
MW	megawatt (electrical)
MWh	megawatt-hour (electrical)
N/A	not available
PRV	pressure-reducing valve
psi	pounds per square inch pressure
STG	steam turbo generator
t/h	tons per hour
WB	wet bulb (air temperature)
/h	per hour
/y	per year

APPENDIX G
Dictionary

Example of a universal system of symbolic names required for programming the unified method for energy systems. Alphabetic classification of names helps to organize the information they contain. This dictionary is by no means complete. Header files of source code contain a particular dictionary for each example computer program.

A

Apart from the ABCD coefficients, names beginning with A contain information associated with a static calculation.

A ABCD coefficient of the performance curve for a unit of plant of index [NN][NU]

AG Generalized ABCD coefficient for all units in a section of plant of index [NN]

AQS Rate of supplied transfer of energy to a section of plant of index [NN] computed by a static calculation

B

Apart from the ABCD coefficients, names beginning with B contain information associated with a progressive best solution.

B ABCD coefficient of the performance curve for a unit of plant of index [NN][NU]

BG Generalized ABCD coefficient for all units in a section of plant of index [NN]

BQS Rate of supplied transfer of energy to a section of plant of index [NN] recorded as part of a progressive best solution

C

Apart from the ABCD coefficients, names beginning with C contain information associated with conversion factors.

C ABCD coefficient of the performance curve for a unit of plant of index [NN][NU]

CG Generalized ABCD coefficient for all units in a section of plant of index [NN]

CVD Conversion factor for a demand of index [ND]

CVN Conversion factor for a section of plant of index [NN]

D

Apart from the ABCD coefficients, names beginning with D contain information about densities of materials, particularly working fluids.

D ABCD coefficient of the performance curve for a unit of plant of index [NN][NU]

DF Density of the final state of material in the processes of a unit of plant of index [NN][NU]

DG Generalized ABCD coefficient for all units in a section of plant of index [NN]

DI Density of the initial state of a material in the processes of a unit of plant of index [NN][NU]

E

Apart from the EFGH coefficients and container EFFY, names beginning with E contain information that has been integrated through an extended period.

E EFGH coefficient of the rate of auxiliary transfer of energy (minor auxiliaries) for a unit of plant of index [NN][NU]

EFFY Name of container. Also, the current value of efficiency or coefficient of performance for the processes and the plant of index [NN][NU], derived from information carried by that container.

EG Generalized EFGH coefficient for all units in a section of plant of index [NN]

EQD Integrated value of the delivery to a demand of index [ND]

EQK Value of energy-in-store at and indexed for the unit time, hour ending IH

EQS Integrated value of the supplied transfer of energy to a section of plant of index [NN]

F

Apart from the EFGH coefficients, names beginning with F contain information about the rates of flow of materials, particularly the working fluids of energy systems. They also contain information about the rates of sources of energy and demands when expressed in their common physical or commercial terms rather than as transfers of energy.

F EFGH coefficient of the rate of auxiliary transfer of energy (minor auxiliaries) for a unit of plant of index [NN][NU]

FG Generalized EFGH coefficient for all units in a section of plant of index [NN]

FMU Rate of flow of a working fluid through a unit of plant of index [NN][NU]

FOIL Rate of flow of oil fuel

G

Apart from the EFGH coefficients, names beginning with G contain information associated with the location of an energy system. In addition, character G is appended to other names, such as AG, to signify that they contain generalized information.

G EFGH coefficient of the rate of auxiliary transfer of energy (minor auxiliaries) for a unit of plant of index [NN][NU]

GALT Altitude of the location of an energy system

GLAT Latitude of the location

GLON Longitude of the location

H

Apart from the EFGH coefficients, names beginning with H contain information about the specific energy of materials and fuels.

H EFGH coefficient of the rate of auxiliary transfer of energy (minor auxiliaries) for a unit of plant of index [NN][NU]

HF Specific energy of the final state of material in a process of a unit of plant of index [NN][NU]

HI Specific energy of the initial state of a material in a process of a unit of plant of index [NN][NU]

HOIL Specific energy (heating value) of oil fuel

I

Names beginning with I contain integer values, particularly for information about time, and they may be used as indices.

I Unit time

ID Day of the month

IE End of extended period

IH Hour of the day (hour ending)

IM Month of the year

J

Names beginning with J contain integer values, and they may be used as indices.

J Index of any one of several interconnected energy systems.

K

Names beginning with K contain integer values, and they may be used as indices. In addition, character K is appended to other names, such as SIZEK, to signify that they contain information about energy storage.

L

Names beginning with L contain integer values.

LCN Number of years (period) for which the lifecycle cost of an energy system is to be evaluated

LCR Number of years (period) of life of an energy system, after which it is expected to require replacement

M

Apart from container MM, names beginning with M contain integer values, and they are generally reserved for management of the computer program. Such names also may be used as indices.

MDEM Logical indicator (TRUE or FALSE) that a computer program is to be used in demonstration mode

MM Name of container

N

Names beginning with N contain integer values, and they may be used as indices. As an example, see the list of values for indices at the end of this Dictionary. Names beginning NB are associated with a progressive best solution.

NAVL Number of units currently available in a section of plant of index [NN]

ND Index of a demand, assigned value

NE Index of a source-of-energy, assigned value

NIMP Number of units imposed in a section of plant of index [NN]

NK Index of a section of plant for energy storage, a particular value of index NN

NN Index of a section of plant, assigned value

NR Index of a residual transfer-of-energy, assigned value

NT Index of a tariff for a source-of-energy, assigned value

NU Index of a particular unit in a section of plant of index [NN], assigned value

NV Index of a step of synthesis representing a particular set of imposed conditions and plant

O

Names beginning with O are reserved.

P

Names beginning with P contain information associated with container PP and about pressures generally.

PA Barometric pressure of ambient air

PAZ Barometric pressure, standard condition

PF Final pressure of the working fluid of a process in a unit of plant of index [NN][NU]

PFMAX Maximum permissible limit for PF

PFMIN Minimum permissible limit for PF

PI Initial pressure of the working fluid of a process in a unit of plant of index [NN][NU]

PP Name of container

Q

Names beginning with Q contain information about transfers of energy.

(Q) The set of transfers of energy accompanying a cell of an energy system

QA Rate of transfer of energy supplied to minor auxiliaries associated with a section of plant of index [NN]

QAU Rate of transfer of energy supplied to minor auxiliaries associated with a unit of plant of index [NN][NU]

QD Delivery to a demand of index [ND] expressed as a rate of transfer of energy

QE Supply from a source of energy of index [NE] expressed as a rate of transfer of energy

QN Rate of needed transfer of energy for a section of plant of index [NN]

QNU Rate of needed transfer of energy for a unit of plant of index [NN][NU]

QP Net sum of the rates of parasitic transfers of energy for a section of plant of index [NN]

QPU Net sum of the rates of parasitic transfers of energy for a unit of plant of index [NN][NU]

QR Rate of residual transfer of energy for a section of plant of index [NN]

QRU Rate of residual transfer of energy for a unit of plant of index [NN][NU]

QS Rate of supplied transfer of energy for a section of plant of index [NN]

QSU Rate of supplied transfer of energy for a unit of plant of index [NN][NU]

R

Names beginning with R contain information about ratios.

RZ Load ratio at a point of discontinuity of the performance curve for a unit of plant of index [NN][NU] at standard conditions

S

Names beginning with SZ contain information associated with container SIZE, about the ratings and loads of plant, all expressed as rates of transfer of energy. They also contain information about the physical size of energy storage.

SIZE Name of container

SIZEK Required physical size of energy storage, evolved as the maximum value of SZK

SZAVL Total available rating of the units in a section of plant of index [NN]

SZHI Maximum permissible rating of a unit of plant of index [NN][NU] at current conditions

SZHZ Maximum continuous rating of a unit of plant of index [NN][NU] at standard conditions

SZK Physical level of energy storage at and indexed for the hour ending IH

SZLO Minimum rating of a unit of plant of index [NN][NU], the lowest load at which it can work continuously

SZOP Current load on a unit of plant of index [NN][NU]

T

Names beginning with T contain information associated with container TT and about temperatures generally.

TADB Temperature of ambient air, dry bulb

TADBZ Air temperature, dry bulb, standard condition

TF Final temperature of a process in a unit of plant of index [NN][NU]

TFMAX Maximum permissible limit for TF

TFMIN Minimum permissible limit for TF

TI Initial temperature of a process in a unit of plant of index [NN][NU]

TT Name of container

U

Character U is appended to other names, such as QSU and QNU, to signify that they contain information about a particular unit of plant of index [NU].

V

Names beginning with V contain information associated with container VV, for costs, and for the evaluation of other attributes and consequences of energy systems.

VC Capital cost of the purchase, installation, and commissioning of a unit of plant of index [NN][NU], dollars per unit of rating

VCG Generalized capital cost of units in a section of plant of index [NN], dollars per unit of rating

VE Cost of supply from a source of energy of index [NE]

VF Annual fixed cost of a unit of plant of index [NN][NU], dollars per year per unit of rating

VFG Generalized annual fixed cost of units in a section of plant of index [NN], dollars per year per unit of rating

VLCE Differential annual rate of escalation of the cost of energy in relation to the rate of inflation, for lifecycle costing, expressed as a decimal fraction

VLCI Discount rate, as the real annual rate of interest after the effect of inflation is removed, for lifecycle costing, expressed as a decimal fraction

VR Running cost of a unit of plant of index [NN][NU], dollars per hour of use per unit of rating

VRG Generalized running cost of units in a section of plant of index [NN], dollars per hour of use per unit of rating

VV Name of container

W

Names beginning with W contain literal strings of characters.

WD Name of a demand of index [ND], including the name of its units of measurement

WE Name of a source of energy of index [NE], including the name of its units of measurement

WM Name of the material in a section of plant of index [NN]

WN Name of a section of plant of index [NN]

WT Name of a tariff of index [NT]

WU Name of the make and model of a unit of plant of index [NN][NU]

X, Y

Names beginning with X and Y generally are reserved for the computer program.

Z

Character Z is appended to other names, such as RZ, SZHZ, and TADBZ, to signify that they contain information associated with standard conditions.

Values for Indices of Symbolic Names

Numerical values of indices used for the examples in this book.

NE sources of energy:

1 Gas fuel
2 Electricity
3 Diesel oil
4 Solar radiation

ND demands:

1 HP process steam
2 Electricity
3 LP process steam
4 Compressed air
5 Hot water
6 Chilled water
7 Heated air

NR residuals:

1 Cooling water

NN sections of plant:

1 LP steam dump valves
2 D/G exhaust water heaters
3 HP/LP steam PRV
4 Refrigeration engine water heaters
5 Deaerators
7 Cold (water/ice) storage
8 Electricity, tie to grid, import
9 Hot (water) storage
11 Cooling water/towers
12 Water/air heaters
15 Electric motor refrigeration

16 Feed-water pumps
17 Diesel generators
18 Backpressure STG
19 Fuel-fired water heaters
20 Absorption heaters (fired)
25 Engine refrigeration
29 Absorption refrigeration (fuel-fired)
33 Electricity, tie to grid, export
34 HP steam boilers
36 Steam/water heaters
40 Absorption refrigeration (steam)
44 Gas turbo generators
46 GTGX HP boilers (unfired)
48 GTGX LP heaters (unfired)
50 Solar water heaters

WARNING: *See the header files of source code for a list of values of indices used by each example computer program, and note that programs may use additional "substitute" values for internal purposes as defined in those lists.*

APPENDIX H

Bibliography

Bridgman, P. W. 1943. *The Nature of Thermodynamics*. Harvard University Press, Cambridge, MA.

Fuller, S. K., and S. R. Petersen. 1995. *Life-Cycle Costing Manual for the Federal Energy Management Program* (NIST HB 135). National Institute of Standards and Technology, Gaithersburg, MD.

Stoecker, W. F. 1989. *Design of Thermal Systems*, 3rd ed. McGraw-Hill, New York.

Tostevin, G. M. 2006. "Computer Synthesis for Energy Plant." Paper 19, presented at the 5th International Conference on Advanced Engineering Design, AED2006, Prague, Czech Republic.

Tostevin, G. M. 2007. "Concept Design of Energy Plant." *Journal of the Energy Institute* 80(2):120–122. Maney Publishing, London.

Tostevin, G. M., and R. E. Luxton. 1979. "The Nature of Thermal Energy Systems." *Mechanical Engineering Transactions of the Institution of Engineers Australia*, ME4:1–10.

Tostevin, G. M., and R. E. Luxton. 1975. "Solar Energy in Industrial Thermal Energy Systems." Paper presented at the Conference on Applied Thermodynamics, Institution of Engineers, Brisbane, Australia.

Tostevin, G. M., and J. C. Nealy. 2003. "Computer Synthesis for the Design of Marine Power Plants." In *Proceedings of the Institute of Marine Engineering, Science and Technology*, B5:25–32. London.

Tostevin, Mark. 1996. *Energy Plants: Design, Operation, Information and Optimization*. Published by the author, Adelaide, Australia.

Installation Instructions for Software Collection

The collection is arranged as a local disk–based website named "EnergySystems," built with Microsoft Expression Web for Microsoft Internet Explorer. It contains computer programs and source code for the demonstrations described in Chapter 6 and Appendix D. The collection also includes users' instructions for those demonstrations, a "Catalog of Examples," "Notes for Programmers," warnings, and notices.

Download the collection from the website associated with this book, www.mhprofessional.com/TostevinEnergySystems. Users' computers need the following:

- A web browser such as Internet Explorer or similar, to read the pages of the collection.
- The Microsoft Windows operating system to run the executable example programs. They are platform-specific and do not run with other operating systems.
- Microsoft Visual Studio, with C++ and FORTRAN, to compile source files into executable programs. However, the source files may also be read and printed with a text editor such as Microsoft Word. See "Notes for Programmers" in the collection.

Software Collection

Download to your computer the folder on the associated website that contains just two items: the folder EnergySystems and the file README.htm. Extract all files if they are .zip compressed. Then double-click README.htm, which will open in your browser. Read the text, and then click the words "Click here" to open the home page of the local website "EnergySystems." Read all pages of the local website for further instructions. See also "Executable Programs" in Appendix D.

Warnings apply to the use of this collection of software. See "No Warranty" and "Warnings" in Appendix D.

Index